RAINBOWS, HALOS, AND GLORIES

Rainbows, halos, and glories

ROBERT GREENLER

Professor of Physics
University of Wisconsin – Milwaukee

CAMBRIDGE UNIVERSITY PRESS

CAMBRIDGE
LONDON   NEW YORK   NEW ROCHELLE
MELBOURNE   SYDNEY

Published by the Press Syndicate of the University of Cambridge
The Pitt Building, Trumpington Street, Cambridge CB2 1RP
32 East 57th Street, New York, NY 10022, USA
296 Beaconsfield Parade, Middle Park, Melbourne 3206, Australia

First published 1980

Printed in the United States of America
Typeset by Vail-Ballou Press, Inc., Binghamton, N.Y.
Printed and bound by Halliday Lithograph Corp., West Hanover, Mass.

Library of Congress Cataloging in Publication Data
Greenler, Robert, 1929–
    Rainbows, halos, and glories.
    Includes bibliographical references and index.
    1. Meteorological optics.   2. Rainbow.   3. Halos
(Meteorology)   I. Title.   II. Title: Glories.
QC975.2.G73   551.5'6   80–143722
ISBN 0 521 23605 3

Dallas Greenler helped me to view the world of nature with an inquiring mind. John Strong helped me develop the tools of science, to enhance my perception of the world. To them, who taught me the pleasures of seeing with the mind as well as with the eye, I dedicate this book.

Contents

Preface

"Whoever sets pen to paper writes of himself, whether knowingly or not." I suspect that E. B. White's assertion is right. In this book I have chosen, knowingly, to include myself; to include, along with my explanation of the rainbow, my personal delight in seeing it; to share not only the results of my investigations of ice-crystal halos, but also my excitement in the process of discovery.

The beginning of my interest lies in my childhood awe of the beauty of the rainbow. My response to something that I like is to try to personalize it by my own participation. Photographing these effects is an attempt to possess them, I suppose, and trying to understand their origins is one of my forms of personal participation. Keats, in his poem *Lamia,* has dealt with the relationship between such understanding and the capacity to appreciate:

> There was an awful rainbow once in heaven:
> We know her woof, her texture; she is given
> In the dull catalogue of common things.
> Philosophy will clip an Angel's wings,
> Conquer all mysteries by rule and line,
> Empty the haunted air, and gnomed mine –

He states, beautifully, an attitude with which I most strongly disagree. For me, understanding enhances the sense of appreciation and wonder. Were I to rewrite Keats, I would, perhaps, end:

> To understand and so become aware
> And, thus, mine beauty from the crystaled air.

With this book I invite you to invest yourself in further understanding.

The book is also about seeing. The world is full of fascinating things that most of us have never seen – obvious things that exist before our eyes, but that we never see. Again and again I am impressed with our blindness to things, however obvious, that we do not already know of. But the rare trait of looking at the world with fresh eyes is one, I believe, that can be fostered and developed. A person who increases the number of areas in which he or she can perceive new things has a source of excitement and satisfaction denied those who do not. This book describes beautiful things that can be seen in the sky, things that can be seen without special equipment or special location, things that can be seen by anyone who sees. I hope that you who read my book will become sensitized to this wealth of sky effects.

My studies of ice-crystal effects (Chapters 2, 3, and 4) have been ongoing over the past dozen years. Much of the work has been done by undergraduate students who became involved and excited by the search for new understanding. Those who have made significant contributions are coauthors of some of the papers listed in the references. However, I want to mention specifically two people who have been major contributors. Jim Mueller became involved as an undergraduate student and continued working with me after he had gone elsewhere to graduate school. Jim Mallmann worked with me on a master's research project that had nothing to do with sky effects; after he graduated he worked with me on understanding sky effects in a fruitful collaboration that has lasted ten years. These people share the credit for many of the results that I refer to as "our work."

Several of the effects illustrated in this book were photographed by me and by others in Antarctica or in the Arctic. Those regions are rich sources of ice-crystal displays that may be seen only rarely in other parts of the world. If we can understand the origins of the optical sky effects, we may use their occurrence and appearance to gain information about atmospheric processes fundamental to the large-scale air movements over the entire earth. This is one of the interests of the Division of Polar Programs of the National Science Foundation, who made it possible for me to go to the ends of the earth to study these effects. I acknowledge their support with thanks.

R.G.

I

Rainbows

The written and oral histories of almost every culture contain myths and legends about the rainbow and its connection with the lives of the people. These legends attempt to answer two kinds of universal questions: The first concerns who we are, where we came from, and how we got to be the way we are; the second concerns the nature of this magnificent arch of colors. This book is my attempt to deal with the second kind of question as it arises in connection with rainbows and other naturally occurring wonders of the sky. If I were to ask why it is that any of us cares about such things, I would, of course, be led back to the first kind of question. Although many of the legends are beautiful and ingenious stories, they do not satisfy my desire to understand more about the real nature of the rainbow.

DESCARTES'S THEORY OF THE RAINBOW

The first person to give a satisafactory explanation, according to my own lights, was René Descartes, in a treatise published in 1637. Many of the earlier speculators on the origin of the rainbow knew that it appeared when sunlight fell on rain drops. Descartes realized that the appearance of the rainbow did not depend critically on the size of the drops and so decided to investigate light passing through one very large drop of water, actually, a spherical glass flask filled with water. His experiment, together with an accompanying theoretical investigation, led him to conclude that the bright (primary) rainbow (Plate 1-1) results from sunlight that enters a spherical droplet of water, is reflected once inside the drop, and then emerges. Sometimes the primary bow is accompanied by a fainter, secondary bow, seen outside the primary; he attributed this bow to light rays that enter the drop and undergo two internal reflections before leaving.

This was more than mere speculation. It is worth going through Descartes's argument in some detail, for it will help us to develop an insight into the rainbow. The approach he used is illustrated in Fig-

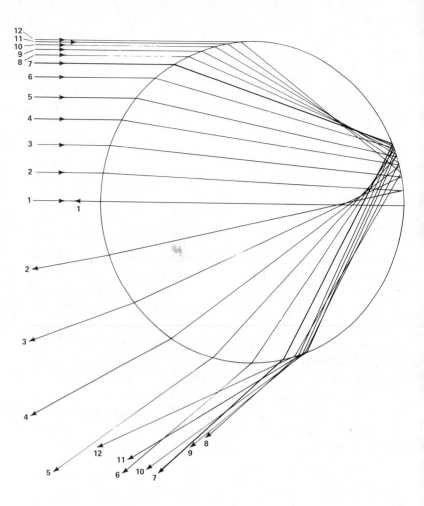

Figure 1-1. Path of light rays through a water drop.

ure 1-1. He mathematically traced a number of parallel rays (coming from the distant sun) through the spherical drop of water. To do this he had to know how a ray of light was deviated from its original direction (refracted when it crossed the boundary from air to water. Willebrord Snell, a Dutch scientist, had discovered the mathematical law of refraction sixteen years before Descartes published his treatise. Snell had not published his work, however, and in the meantime he had died. Descartes's published work included a derivation of the law of refraction presented as his own, which led to a wrangle within the scientific community. Although Descartes disputed the charge, some people claimed that he had seen Snell's manuscript and had appropriated his idea. It is interesting to note that the disputed law of refraction, called Snell's law in most countries, in France is referred to as Descartes's law.

In any case, Descartes knew the law of refraction and therefore could calculate the path of a light ray passing into and out of the drop. Because the law of reflection was also known, he could calculate and trace the complete path of any ray entering the drop and

2

being reflected inside. From Figure 1-1 you can see that the ray striking the drop at its center is reflected directly back along its incident path. Call this path the axis. Then you can see that rays entering the drop above the axis come out of the drop at an angle below the axis. Rays entering the drop farther above the axis emerge at greater angles below, but this trend continues only up to the ray shown on the diagram as a heavy line (the Descartes ray). Rays entering the drop above that heavy line emerge at smaller angles to the axis. The angle of the exit ray (measured to the axis) is a maximum value for the Descartes ray. As a consequence, rays that enter the drop close on either side of the Descartes ray emerge at about the same angle as that ray; that is to say, there is a concentration of rays emerging from the drop at that maximum angle. Descartes concluded that there are more rays emerging between the angles of 41 and 42 degrees than in any other one-degree interval, and it is this *concentration* of rays near the maximum angle that gives rise to the rainbow.

Because of this, in looking in the sky to see the light that enters a water drop and is reflected back out of the drop, we look 42 degrees away from the "straight-back" direction. To understand this it is helpful to define the straight-back direction more carefully. To see rays that come right back along the axis we would look in the direction exactly opposite the sun. This point in space, directly opposite the sun from an observer, is called the antisolar point. If the sun is above the horizon the antisolar point will lie below the horizon, and, in fact, its position will be marked by the shadow of the observer's head on the ground. Figure 1-2 shows the path of a ray of light from the sun that enters a drop and is reflected back out of it at an angle of 42 degrees directly to the eye of the observer. From the figure we can see that the ray that enters the eye also makes a 42-degree angle with the direction of the antisolar point. Once we understand that idea,

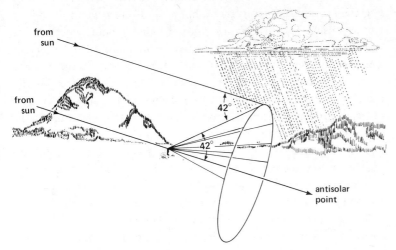

Figure 1-2. From Descartes's construction, the rainbow is predicted to be a circle of angular radius 42 degrees, centered on the antisolar point.

3

the shape and position of the rainbow follow easily. To see the rainbow we look in a direction (any direction) that is 42 degrees away from the antisolar direction. That condition describes a circle around the antisolar point. A convenient way to describe the size of such a circle in space, where the angle between lines pointing to the center and to the edge of the circle is 42 degrees, is to say that the circle has an angular radius of 42 degrees.

So Descartes started from a rather simple hypothesis of how the rainbow was formed. (We often refer to such a hypothesis – light rays entering a spherical water drop, for example – as a model.) He had to use one experimentally determined number, the index of refraction of water, to calculate how much the light was refracted on entering and leaving the drop. But from this model he predicted that the rainbow should lie on a circle, of angular radius 42 degrees, centered on the antisolar point. When he did the same type of calculation for light rays that were reflected twice within the drop, he found that the emerging rays came back at a maximum angle of 51 degrees to the axis. So by the same line of reasoning, his model predicted that the secondary bow should also lie on a circle about the antisolar point, but with a 51-degree angular radius. Both these predictions proved to be correct. Look at Plate 1-2. The photograph was taken with a very wide-angle camera lens. You can see my shadow on the field in the foreground and see that both the primary and secondary bows are circles about the shadow of my head, which marks the antisolar point. Descartes was right. Although we have gained further understanding of rainbows, I find Descartes's explanation quite satisfying.

SIZE OF THE RAINBOW

I have discussed the size of the rainbow in terms of the angle between the center and edge of the circle, a process that may seem to beg the question how big the rainbow is. There is good reason for this approach: The angular size of the bow is the same whether it is formed in the spray of your garden hose, held at arm's length, or in a sheet of rain a few miles away. From Figure 1-2 you can see that the primary rainbow can be formed from the contribution of all the raindrops that lie on the surface of a cone with its apex at the eye, its center along the antisolar direction, and its half angle equal to 42 degrees. In such a case, the size of the rainbow measured in feet or miles has no meaning because there is no specific distance from the observer where the bow can be located. The angular size does have a definite meaning and is the reasonable way to describe, not only the rainbow, but most of the other phenomena I will discuss.

The fact of this constant angle can be demonstrated easily by your frustration in trying to photograph the entire bow. A camera's field

4

of view is defined by an angle (a typical standard lens for a 35-millimeter camera will cover a field about 40 degrees across the wider dimension of the picture). The bow formed in the spray of your hand-held garden hose might seem small enough to photograph without difficulty, but when you look through the camera viewfinder you will see only a small portion of it. Your normal response is to step back to include more of it, but as you step back, the rainbow, which is localized in the hose spray, expands so that you still see only the same small portion of it through your camera; its angular diameter remains constant, although its estimated diameter as you see it localized in the hose spray increases with distance.

When we consider that only raindrops located on the surface of a cone (whose tip is centered on the observer's eye) can contribute to the observed rainbow, another interesting idea comes up. It is clear that two people, standing side by side admiring "the rainbow," are actually seeing light refracted and reflected by different sets of raindrops. Each person has his or her own personal rainbow.

COLOR OF THE RAINBOW

Descartes explained the fundamental features of the rainbow but was unable to explain the presence of the rainbow colors. It was not until some thirty years later that Isaac Newton understood that white light is a mixture of light of all colors and that the index of refraction of water (or any transparent material) is slightly different for light of different colors. Red light gets refracted less than blue in passing from air to water and hence comes out of the droplet in a different direction from the blue light. The extreme angle of the emerging light varies from about 42 degrees for red light at one end of the spectrum to about 40 degrees for violet light at the other end. The bow for each color has a slightly different angular size. The result of the superposition of a continuum of bows of different sizes and different colors is the display of colors with which we are familiar. Note that in the primary bow the red is on the outside of the bow, whereas the color sequence is reversed in the secondary bow.

HIGHER-ORDER RAINBOWS

Each spherical raindrop reflects and refracts, in all possible ways, the light that strikes it. A fraction of the light is reflected when the light first strikes the drop; the rest is transmitted through the drop until it strikes the droplet surface from the inside. Here again a fraction is reflected, remaining in the drop, and the remainder leaves the drop. Thus, at each encounter with the surface inside the drop, a portion of the light is reflected and remains in the drop and the rest escapes. From a ray entering the drop at any particular point, rays

5

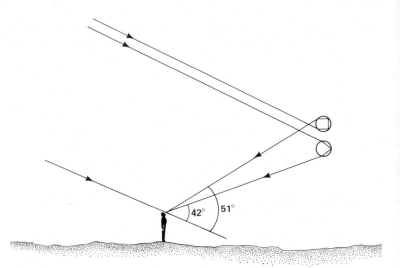

Figure 1-3. Different raindrops contribute to the primary and to the larger, secondary rainbow.

can emerge after one, two, three, or more internal reflections, and the paths and emerging angles will be different for the different colors contained in the sunlight. This complicated situation is simplified by the understanding that, for a particular rainbow, you can "see" only the particular ray that comes to your eye. Thus, Figure 1-3 shows two drops, one contributing to the primary bow and the other to the secondary. Both drops have singly and doubly reflected rays, but the diagram shows only the rays that go to the observer's eye.

What about rainbows caused by more internal reflections? In the absence of any detailed understanding of how the rainbows are produced, you might expect to see a fainter tertiary bow outside the secondary, a yet fainter quaternary bow, and so on. Apparently, Descartes never extended his method to rays experiencing three internal reflections. One of Newton's many significant accomplishments was his development of a powerful mathematical method, the calculus, which turned out to be a beautiful tool for doing, simply, what Descartes had done by a long and tedious series of calculations. Using calculus, Newton derived a mathematical expression for the angular size of the rainbow after an arbitrary number (N) of reflections. For $N = 1$, his expression gave about 42 degrees, and for $N = 2$, about 51 degrees. (I say "about 42 degrees" because the precise value depends on the value used for the index of refraction, that is, on the color we are considering.) It is curious that nowhere in his writing does Newton refer to having put the value $N = 3$ in his equation. In his book *Opticks*[1] he states, without further comment, "The Light which passes through a drop of Rain after two Refractions and three or more Reflexions is scarce strong enough to form a sensible bow." Boyer,[2] in his excellent book on the history of the rainbow, quotes Newton's contemporary Jean Bernoulli, suggesting that the third

bow might be visible to eagles or lynxes but not to human eyes.
Boyer goes on to describe the involvement of Edmund Halley (of
comet fame) with the rainbow problem:

> Halley seems to have been the earliest to carry through to the
> end the calculations on the tertiary rainbow; and the result
> must have been a surprise. He found that the third rainbow arc
> has an angular radius of 40° 20′, and that it should appear not
> in the part of the heavens opposite to the sun, but as a circle
> *around the sun itself*. For at least two thousand years men had
> been looking for this arc in the wrong part of the sky!

The reason we do not see this bow may not actually be its faintness,
as suggested by Newton and Bernoulli, but rather its appearance
near the sun where the general sky background is much brighter
than on the opposite side of the sky, so that it is swamped by the sky
light. The fourth-order bow should also be a circle around the sun,
with an angular radius of 46 degrees, and not until we consider rays
undergoing five internal reflections do we expect a bow in the sky
away from the sun.

After five reflections it may be that the light is too faint to be visible even to Bernoulli's eagles or lynxes. In the laboratory, Walker[3]
has observed, in a single drop of water illuminated with a laser
beam, the emerging rays corresponding to all of the rainbow orders
up to the thirteenth and found them to be located in positions
closely agreeing with the rainbow model of Descartes. I do not
believe that any of the bows of higher order than the secondary have
been observed in nature, however, although a number of people have
told me that they have seen three rainbows. Some of these observations may involve a portion of a reflection bow and some may involve
interference bows (both of which will be discussed later). And some
involve the kind of faulty observation to which most of us are more
prone than we realize. However, some of the reports seem to be so
detailed that I wonder if there occasionally appears a bow that we do
not understand.

THE BRIGHTNESS OF THE SKY NEAR A RAINBOW

From Descartes's drawing (Figure 1-1) we can understand another
interesting feature of rainbows. Remember that the rainbow is
caused by the concentration of rays emerging from the drop at the
extreme angle. Light also emerges from the droplet, however, at
smaller angles. To see a ray coming straight back out of a drop you
would look at the antisolar point. To see other rays that came out at
a small angle to the straight-back direction you would look at a
small angle from the antisolar point. All of these "extraneous" rays
that get to the eye should come from drops located on the inside of

the primary bow. Now look again at some of the rainbow photos (or better yet, at some rainbows). The sky background is brighter inside the bow than outside, as you can see in Plates 1-1 and 1-2. Seldom is the effect as dramatic as in the magnificent picture of Plate 1-3, but it is almost always visible. It is a standard feature of rainbows, as much a part of the rainbow phenomenon as the colored arc.

Plate 1-4 is a photograph taken with an inexpensive, plastic-lensed camera. It shows the brighter sky inside the portion of the bow that stands in front of the mountain. But it also shows a feature that, in my experience, is relatively rare. If we do Descartes's ray-tracing trick for the secondary bow (two internal reflections), we find that the "extraneous" rays, which do not contribute to the secondary arc, should be seen *outside* the secondary. The effect can be seen in Plate 1-4, where the sky is brighter inside the primary and outside the secondary than it is between the two bows.

INTERFERENCE (OR SUPERNUMERARY) BOWS

In some of the rainbow photographs already discussed, you can see an additional faint arc or series of arcs just inside the primary bow,

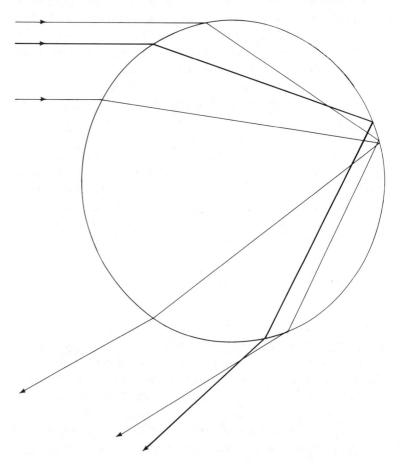

Figure 1-4. A pair of rays that can contribute to an interference bow inside the primary rainbow.

usually more obvious near the top of the bow than at its base. These result from the wave nature of light and are called interference bows or supernumerary bows. Again, Descartes's drawing affords a good starting place for understanding. In Figure 1-4 the diagram of Figure 1-1 is redrawn to show only three of the rays. The dark ray is the Descartes ray, the one emerging at the extreme angle. The other two rays shown are chosen from among those entering the drop on either side of the Descartes ray; after taking different paths through the drop they emerge, going *in the same direction* to the eye of an observer. Let us consider the consequences of the fact that light has the properties of a wave. The waves corresponding to the rays entering the drop are all in phase (i.e., in step with each other, their crests and troughs lined up side by side). The two rays that come out of the drop parallel to each other, however, have traveled paths of different lengths in passing through the drop. Suppose that the difference in the path length for the two rays is just equal to 0.5 wavelength. In such a case the rays coming out of the drop will be out of phase (out of step with each other, so that the crest of one wave matches the trough of the other). In this situation the waves will cancel each other — will mutually annihilate themselves — with the result that a person looking in that particular direction sees a reduced light intensity.

A pair of rays entering the drop more widely separated from the Descartes ray will emerge at a smaller angle (measured to the axis). If I pick the pair of rays whose paths through the drop differ by one complete wavelength, they will again be in phase and will reinforce each other, giving an intensification of light emerging from the drop at that angle. As I continue to pick successively more widely spaced incident rays, I will find pairs whose path differences are 1.5 wavelengths (producing a diminished light intensity), 2 wavelengths (enhanced light intensity), 2.5 wavelengths (diminished), and so on. The net effect of the interferences of these pairs of light beams as seen by an observer is a series of light and dark bands just inside the primary rainbow.

The interference of light beams to produce an enhancement or diminution of light intensity is not unique to the rainbow. The effect can be shown in many different experiments — experiments that, by showing constructive and destructive interference, have convinced us of the wave nature of light.

The spacing of these interference or supernumerary bows depends on the size of the droplets: Small drops yield a larger spacing than larger drops. You may see up to three or four supernumeraries, but frequently you see none. If a wide range of drop sizes is present, the interference bows with different spacings overlap to cancel out the clear variation of intensity that results from droplets all of the same size. In my observation, the supernumeraries do not show bright spectral colors but usually seem to alternate between a pale blue or violet and a pale green.

THE BRIGHTNESS OF THE RAINBOW
NEAR THE GROUND

Fraser[4] has presented an interesting idea that seems to explain two commonly observed features of the rainbow. When the sun is near the horizon, the ends of a rainbow near the ground are frequently much brighter than the rest of the bow. The supernumeraries for such a bow are generally more obvious higher up on the bow and disappear near the ground. Fraser explained both of these features as resulting from the presence of a mixture of large and small water drops. The surface tension of water acts as a tight skin over a water droplet, tending to squeeze it into a spherical shape, and this tension is sufficient to keep a drop smaller than a few tenths of a millimeter quite spherical as it falls through the air. A large falling drop, however, is distorted from a spherical shape by the air-drag forces. Such a drop assumes a squashed, round-pillow shape, with a circular cross section in a horizontal plane but a reduced vertical dimension. (Note that this is quite different from the stereotyped cartoon representation of a teardrop shape with a vertical tail.) When the sun is near the horizon, the rays that would form the foot of the rainbow travel through the large drops, on their way to the observer's eye, in a horizontal plane. In this plane the drops have circular cross sections that produce a rainbow. Because they are large drops they may contribute significantly to the intensity.

It is a different matter for the rays that would form the top of the rainbow. On their way to the eye of the observer, they travel through vertical cross sections, which are not circular and hence cannot contribute to the rainbow intensity.

We see that all sizes of water drops can contribute to the portion of the rainbow near the horizon, but each different size of drop contributes to a series of supernumerary bows of different spacing. So, near the horizon, the wide range of drop sizes gives a bright bow but blurs out any interference structure. Higher up on the bow, only the smaller drops are contributing to the bow, and those small drops may be close enough to being the same size to give the supernumerary structure.

THE WHITE RAINBOW

Up to this point I have discussed the rainbow mostly in terms of ray optics, where we assume that light travels in straight lines, described adequately by the laws of refraction and reflection. I did invoke the wave nature of light to explain the supernumerary bows. A more sophisticated treatment of the rainbow would have to take account of another manifestation of the wave nature of light: diffraction. If light from a small source falls on a piece of cardboard with a circular hole in it, a circular spot of light will be seen on the wall beyond the cardboard, and the spot of light will get smaller as the hole is made smaller. This is part of our everyday experience. However, as the hole is further reduced in size until it is very small, there comes a point when the spot becomes fuzzy, without a sharp edge, and then actually increases in size as the hole is yet further reduced. This spreading of the light beam when it passes through a very small aperture is called diffraction and is a well-understood feature in the propagation of any kind of wave. To produce diffraction spreading, the light does not have to pass through a hole but can pass through anything that limits the width of the beam – like a very small lens, or a very small water drop. The diffraction effect of rainbow rays passing through a very small water droplet is to broaden the angular band of the emerging green light, for example, so that it overlaps the broadened bands of red and of blue light. In the extreme case where all of the colors overlap, the result should be a broad white rainbow.

Plate 1-5 shows a white rainbow, sometimes called a cloud bow or fog bow, because clouds and fogs are typically made up of the very small drops that produce this effect. After I had taken this photograph, I looked at it several times over the succeeding months. There was something about it that was vaguely bothersome until I recognized that inside the broad white bow was another faint bow. It may be difficult to see on the reproduction of Plate 1-5, but, after I recognized it, it was obvious on the original slide; indeed, it was obvious to anyone to whom I pointed it out. It is yet another of many examples of blindness to things we do not understand. In trying to identify this extra bow, several degrees inside the bright white bow, I considered whether it could be a supernumerary bow (remember that the spacing of the supernumerary bows depends upon the drop size and is larger for smaller drops). I turned the question around and asked: If this extra bow is the first supernumerary, spaced so far from the primary, what would be the diameter of the droplets necessary to produce it? The calculation produced the answer that the droplets would have to be about 10 micrometers (10 millionths of a meter) in diameter, which is typical droplet size for clouds or fogs. So it appears that the white rainbow has a supernumerary companion.

I see the white rainbow sometimes while flying in an airplane over a smooth, featureless deck of clouds. The way to look for it is to locate the antisolar point by seeing from what direction the sun is shining into the airplane windows. In such a case the antisolar point may well be surrounded by colored rings (the glory), to be discussed in Chapter 6. About 40 degrees away from this point look for a band

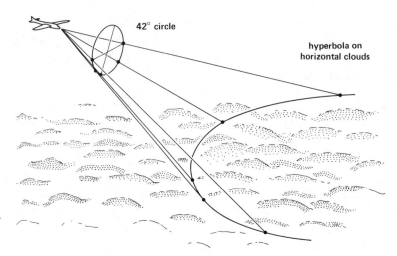

Figure 1-5. A rainbow localized on a horizontal cloud layer has the shape of a hyperbola.

of white. In the case of a well-defined cloud deck, you get the clear impression of the white rainbow localized on the surface of this horizontal layer. The figure of this horizontal rainbow is actually the intersection of the rainbow cone (of half angle about 42 degrees; see Figure 1-2) with the horizontal plane. The illustration in Figure 1-5 shows the case where this intersection has the form of a hyperbola. If the sun elevation were greater than 42 degrees, the figure would be an ellipse and would turn into a circle only if the sun were directly overhead. Those of us who look for such effects from an airplane are frustrated by the restricted field of view afforded by the windows of most commercial planes. We are seldom able to see more than a portion of the horizontal cloud bow, even when it is all present. Plate 1-6 shows a portion of the cloud bow viewed from an airplane.

THE RED RAINBOW

Plate 1-7 is a photograph of the red rainbow. Note that the rainbow is high in the sky, which implies that the sun is close to the horizon on the other side of the sky. The bow is formed by the small water droplets that make up the clouds and would have been white except for one effect: Because the sun is almost setting, the sun's rays travel through a long path in the earth's atmosphere before reaching the clouds in the picture. As will be discussed later, sunlight going through such a long atmospheric path selectively loses the blue end of the light spectrum by scattering. The white sunlight with the blue part of the spectrum partially removed is the reddish color you see in the red cloud bow.

THE RISING RAINBOW

The sequence of three photographs in Plate 1-8 was taken off the coast of Hawaii. The photographer first noted the red band near the horizon in the middle of the afternoon and, not knowing what caused it, took a picture. He took the other two pictures at 10- or 15-minute intervals and brought them to me for an explanation. He was seeing the rising rainbow. In the first picture the antisolar point is about 40 degrees below the horizon, so that just the very top edge of the bow appears above the horizon. This implies that the sun must be about 40 degrees above the horizon on the other side of the sky. As the sun sets, the rainbow rises.

REFLECTED-LIGHT RAINBOWS

Look at the remarkable photograph of Plate 1-9. The picture was taken near the California coast in the evening. Because the sun would be setting in the west, the observer would view the rainbow with the sun and ocean at his back. The ocean must have been very calm that day, because the second set of rainbows was formed by a mirrorlike reflection of sunlight from the ocean surface. Let us consider where we should see the bow resulting from the reflected light. Figure 1-6 shows light coming back out of rain drops at the rainbow

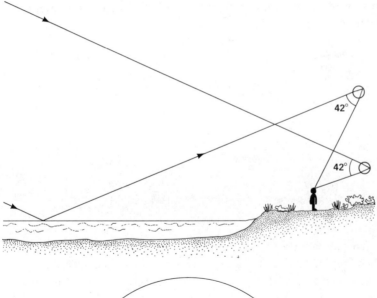

Figure 1-6. Rays contributing to the rainbow and to the reflected-light rainbow.

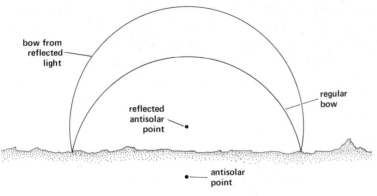

Figure 1-7. The ordinary rainbow is centered on the antisolar point; the reflected-light bow, on a point located at the sun's elevation above the horizon.

13

angle of 42 degrees, both for light striking a drop directly and for light striking a drop after reflection from the ocean surface. From this drawing we can see that the reflected light should produce a bow higher in the sky than the direct light. Figure 1-7 is a drawing of these bows as seen by an observer. To draw a rainbow, all that is needed is to locate the antisolar point and draw a 42-degree circle around it. For the sun above the horizon, the antisolar point will be below the horizon, as shown in Figure 1-7. The reflected light acts as if it were coming from a sun below the horizon. In fact, if you were to turn around and look at the reflected light, you would see the reflected sun below the horizon.

So the antisolar point for the reflected light lies above the horizon, and the reflected-light rainbow should be a circle around it. Clearly, we have the possibility of seeing more than half of the rainbow circle for the reflected-light bow. A little exercise in plane geometry can convince you that the two bows should intersect at the horizon, no matter what the elevation of the sun. For the secondary bows it is necessary only to draw 51-degree circles about the same two points. To see these reflected-light bows, one would not have to be standing in view of the reflecting water surface, which for a low sun might be some miles away.

Figure 1-8. The predicted form of the direct and reflected-light primary and secondary bows. The dotted line is the area covered by the photograph of Plate 1-9.

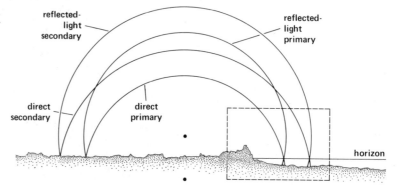

reflected-light secondary

reflected-light primary

direct secondary

direct primary

horizon

To compare the theory with the observation (the photograph of Plate 1-9 in this case), I have drawn the direct and reflected-light rainbows for a sun elevation of 11 degrees (Figure 1-8). The dashed rectangle shows the approximate area covered by the photo, and I would say that the agreement in shape makes the explanation convincing. This effect might even account for some of the reports of three rainbows in the sky.

REFLECTED RAINBOWS

I have referred to rainbows seen above the horizon, resulting from reflection behind the observer, as reflected-light rainbows in order to save the term *reflected rainbow* for the case where one can see a rainbow in the sky and simultaneously see its inverted reflection in a water surface. The reflected rainbow appears to be as ordinary as the reflection of any object, but there is a slight difference. Boyer[5] points out the difference by calling attention to a paragraph from a book of the last century entitled *Budget of Paradoxes,* by Augustus De Morgan: "A few years ago an artist exhibited a picture with a rainbow and its apparent reflection . . . Some started the idea that there could be no reflection of a rainbow; they were right; they inferred that the artist had made a mistake; they were wrong."

That the artist made no mistake can be confirmed by photographs.[6] A key to the paradox lies in the realization discussed earlier, that each observer sees a different rainbow, that is, a rainbow produced by a different set of water drops. The reflected bow is not a direct reflection of the bow seen simultaneously in the sky. Figure 1-9 illustrates that the reflected and direct bows are produced by different sets of drops.

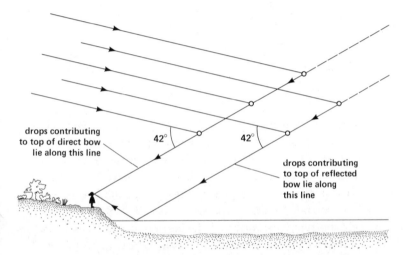

drops contributing
to top of direct bow
lie along this line

42°   42°

drops contributing
to top of reflected
bow lie along
this line

Figure 1-9. The reflected rainbow is produced by a different set of drops from those that produce the regular bow.

RAINBOWS ON WATER SURFACES

There is yet another situation where I have seen reflected-light rainbows in nature. On a few occasions I have been in a canoe on a lake, shrouded by an early morning fog, before sunrise. As the sun rose, the fog cleared rather quickly and I saw a portion of a rainbow, localized on the lake surface. In one case, close examination revealed small droplets of water sliding about on the water surface. Although I can speculate at some length on the conditions that produce water droplets remaining on the surface without merging with the lake water, I do not understand the process. However, given the drops on the smooth water surface, I do understand something about the rainbows they produce. They can give rise to a direct rainbow and to another bow resulting from light reflected off the lake surface, either before or after it passes through the drops.

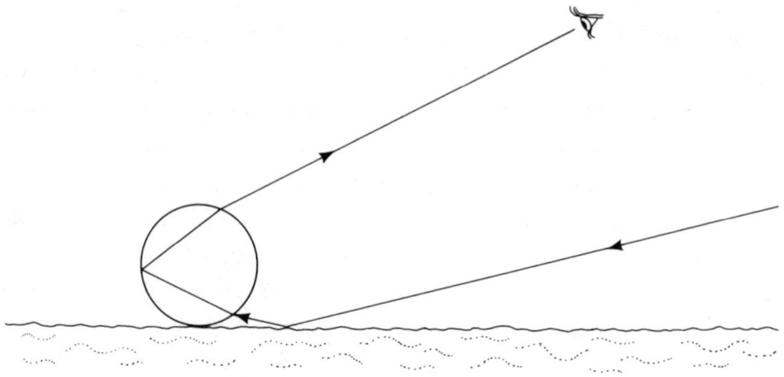

Figure 1-10. A reflected-light rainbow can be formed by drops resting on a reflecting surface.

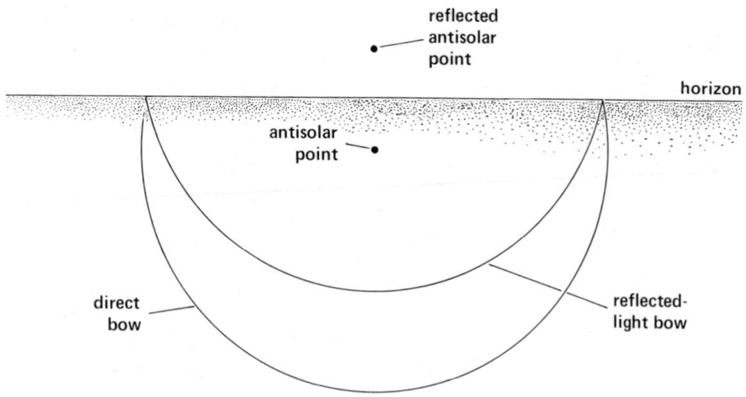

Figure 1-11. The shapes of the direct and reflected-light rainbow from water drops sitting on a water surface.

Figure 1-10 illustrates the case where a ray is reflected before it enters the drop. The effect of the reflection is to move the effective antisolar point for the reflected-light bow above the horizon, and the construction of the two bows is shown in Figure 1-11. Rainbow light that is reflected after passing through the drops will result in bows in the same location. The appearance of the bows is modified by their apparent localization on the horizontal water surface, where for a sufficiently low sun angle they will appear to be hyperbolas. Plate 1-10 shows a reflected-light bow for a high sun elevation.

It is interesting to note the similarity of Figures 1-7 and 1-11. Think of the very special set of circumstances of water drops on a lake surface in front of you and a smooth reflecting body of water behind you, in the direction of the sun. The combination could produce two complete circles for each of the two rainbows, extending above and below the horizon, though I have never heard the effect discussed or seen it.

RAINBOWS EVERYWHERE

Those of us who are hooked on rainbows see them in many places. The sprays from hoses, fountains, or waterfalls (Plate 1-11) are obvious places to look. Other possibilities are the bow wave of a boat, the blown spray of a breaking wave, or spray thrown up by the tires of passing cars on a wet road. You can see rainbows flit through the shower of each lawn sprinkler as you drive down a residential street, and you can see them where they lie trapped in the drops of dew on a lawn (Plate 1-12). I sometimes move so that I can see the individual dew drops hanging from a spider web light up with a rainbow brilliance hardly matched by that of a diamond. When the web is close-knit and has collected a plane of small drops, the entire plane lights up with either the primary or the secondary bow as I move my head.

To position yourself to be able to see the rainbows in drops at a particular location, look for the shadow of your head and move so that it falls 42 degrees away from your rainbow-making drops. You have a fairly convenient angle measurer literally at hand: When I open my hand to spread the thumb and little finger as far apart as possible and hold it palm outward at arm's length, the angular distance between thumb tip and finger tip is about 22 degrees. If I start by placing my outstretched thumb over the shadow of my head, a quick measure of two hand spans tells me where the drops should be to light up with the rainbow colors. Some people have small hands; but smaller hands usually go with shorter arms, and the combination tends to preserve this 22-degree angle.

There are still more places to watch for rainbows. The breast-stroke swimmer, exhaling as he brings his head out of the water, throws up a mist so close to his face that each eye may see a rainbow separately and he has the view of twin rainbows.[7] Bright bands across the beam of a searchlight pointed skyward on a rainy night can be identified with the primary and secondary rainbows:[8] Try to decide whether the primary rainbow is the lower or higher band across the light beam (your first guess will probably be wrong).

LUNAR RAINBOWS

The full moon can act as a source of light for rainbows. Most observers of lunar rainbows report them to be white, but I suspect that this is usually a physiological factor. At low levels of illumination, the eye loses its color sensitivity, so that a standard multihued bow appears white. This effect has been nicely illustrated by a photograph[9] of a lunar rainbow showing the usual colors, although the photographer reports that this bow appeared to be white when he took the picture.

THE INFRARED RAINBOW

I have put considerable research effort into investigating the structure of molecules that become attached to the surface of solids. This attaching process (adsorption) is of considerable importance in a variety of things, such as the functioning of catalysts, the separation of ores, processes within a fusion reactor, and many others.

It would at first seem that this activity should have nothing to do with rainbows, but our assumption here is not quite correct. One of my main tools for looking at the structure of adsorbed molecules has been infrared radiation. If we send white light through a prism it will be spread out into the familiar spectrum of colors that constitute it. The physical difference between red light and blue light, for example, is that the red-light waves are longer than those that are blue. The wavelength of the radiation (or light) becomes a more precise measure of the different parts of the spectrum than the color. So the visible spectrum is made up of waves that are all of the same kind, except for their different wavelengths. In fact, this same spectrum of waves has some whose wavelengths are even longer than the red we can see. These are the infrared waves. Still longer waves of the same kind are microwaves and radio waves. At the other end of the visible spectrum, waves that are shorter than the visible violet are called ultraviolet.

Having worked with infrared radiation, I was familiar with its properties, and so it is perhaps understandable that the question occurred to me: Is there an infrared rainbow in the sky? For there to be an infrared bow, the source of light, the sun, would have to be emitting infrared radiation – which it does. That radiation would have to get through the earth's atmosphere to the water drops. The water vapor and carbon dioxide in the atmosphere absorb some infrared wavelengths, but others are transmitted. Next, the radiation would have to go through a water drop. Liquid water strongly absorbs many infrared wavelengths, but there is still a range of wavelengths, not far from the visible, that should get through a millimeter thickness of water. Some of this radiation should proceed to the eye of the observer, where he or she cannot detect it. Such a train

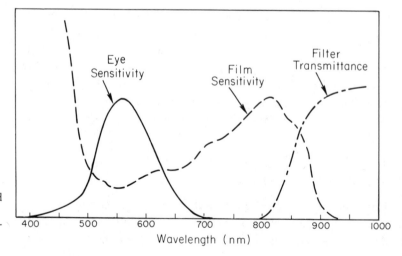

Figure 1-12. The sensitivity of the film and the transmittance of the filter used to photograph the infrared rainbow. The eye sensitivity curve is shown to define the limits of the visible spectrum.

of thought makes the answer to the question clear: Yes indeed, there should be an invisible infrared rainbow in the sky, lying just outside the red band of the primary bow. In searching the scientific literature, I could find no evidence that anyone had actually detected such a rainbow; yet it seemed to me an interesting thing to look for.

My approach was to photograph it, using infrared sensitive film. The essentials of the project are summarized in the graphs of Figure 1-12. The horizontal scale is wavelength. The most fundamental way to define the limits of the visible spectrum is with the curve of the sensitivity of the human eye to different wavelengths, and from that curve in Figure 1-12 we can see that the blue limit is at about 400 nanometers (400 billionths of a meter) and the red at about 700 nanometers. Another of the curves shows that the infrared film is sensitive to about 930 nanometers. The problem in taking a picture with the infrared film is that it is sensitive not only to the infrared but to all of the visible, and it is extremely sensitive to blue light. An unfiltered exposure would result in a black-and-white negative,

Figure 1-13. The infrared rainbow in a water spray. (Photographed by the author)

but there would be no way to tell what part of the spectrum produced what part of the exposure. The key to sorting out that problem is the filter whose transmission characteristics are also shown in Figure 1-12. This filter looks like a sheet of black plastic, opaque to visible light. Its transmission curve shows that it transmits only wavelengths longer than about 800 nanometers. With this film and this filter, the only wavelengths that can contribute to the exposure lie in a band between 800 and 930 nanometers, clearly separated from the visible light.

Figure 1-13 shows one of my first attempts to photograph the invisible bow. On top of the ladder in the foreground is a board supporting a section of hose punctured with many small holes. In the spray from the hose you can clearly see the infrared primary and secondary bows. A couple of supernumeraries can be seen inside the primary, and on the original negative, I can even see a supernumerary outside the secondary bow. (I have never seen that supernumerary in any other rainbow or in photographs of rainbows.) That

Figure 1-14. A natural infrared rainbow. (Photographed by the author)

20

was an exciting start, but it took over two years until I saw a natural rainbow when I had, immediately at hand, my camera *and* the infrared film *and* the filter and took the photograph of Figure 1-14. When the film was developed, I saw for the first time an infrared rainbow that had hung in the sky, undetected, since before the presence of people on this planet.[10]

SINCE DESCARTES

There have been a number of more sophisticated mathematical treatments of the rainbow since Descartes's attempt three and a half centuries ago. In a recent article, Nussenzveig[11] describes some of these theories, including his own contributions of the past few years. Descartes illuminated the physical process by which the rainbow is produced. But the appearance of the rainbow does depend on drop size, a parameter that did not appear in his theory. He had no satisfactory explanation for the colors or the supernumerary bows. Subsequent theories have described the intensity distribution. Nussenzveig's complex mathematical treatment yields a detailed description of the rainbow and also of the glory (the colored rings around the antisolar point); it thus goes far beyond Descartes's theory in its explanatory power.

A RAINBOW PUZZLE

I will close this discussion by giving you a rainbow puzzle to consider. The puzzle is contained in the photograph of Plate 1-13 and is this: Can you decide what is strange about this picture? There is no trick photography; I took the photo because it was unusual. If I provided the answer to this puzzle immediately, some of you who would actually enjoy pondering the question would not be able to stop to consider it before reading the answer. To preserve that pleasure for you, I give the answer in the Appendix at the back of the book.

Plate 1-2. Primary and secondary rainbows, centered on the antisolar point (marked by the shadow of the photographer). (Photographed in Wisconsin by the author)

Plate 1-1. Rainbow. (Photographed in Wisconsin by the author)

Plate 1-3. Rainbow with dramatic background light inside the primary bow. (Photographed in California by Allen L. Laws)

Plate 1-4. Brighter sky inside the primary bow and outside the secondary than between the two bows. (Photographed in Colorado by Daniel G. Grimm)

Plate 1-5. White rainbow, fog bow, or cloud bow with the first supernumerary bow. (Photographed in Ontario by the author)

Plate 1-6. Cloud bow seen from an airplane. (Photographed by Kenneth Sassen)

Plate 1-7. Red rainbow (cloud bow formed by red light from the setting sun). (Photographed in Washington by Alistair B. Fraser)

Plate 1-8. Sequence showing the rising rainbow. (Photographed in Hawaii by William L. Walters)

Plate 1-10. Reflected-light rainbow formed by droplets on the water surface. (Photographed in Wisconsin by the author)

Plate 1-12. Dew bow. (Photographed in Wisconsin by Walt Tape)

Plate 1-9. Primary and secondary rainbows with their reflected-light companions. (Photographed in California by Allen L. Laws)

Plate 1-11. Rainbow in spray from waterfall. (Photographed in New Zealand by Harold G. Muchmore)

Plate 1-13. A rainbow puzzle, explained in the Appendix. (Photographed in New York by the author)

Plate 2-2. Portion of a 22-degree halo. (Photographed in Wisconsin by the author)

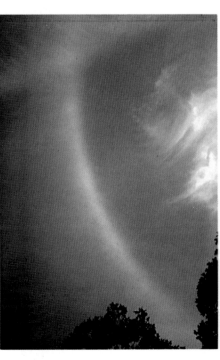

Plate 2-1. Complete 22-degree halo. (Photographed in Wisconsin by the author)

Plate 2-4. Sun dog showing white extension away from the sun. (Photographed in Wisconsin by Harvey J. Lindemann)

Plate 2-5. Sun dog showing spectral colors. (Photographed in Wisconsin by Harvey J. Lindemann)

Plate 2-3. Setting sun with 22-degree parhelion (sun dog). (Photographed in Wisconsin by the author)

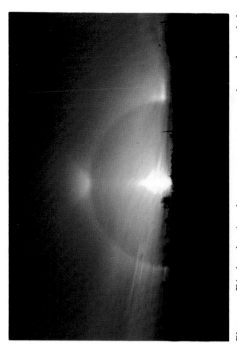

Plate 2-6. Multiple exposure of the sun, taken in Antarctica at the South Pole. (Photographed by the author)

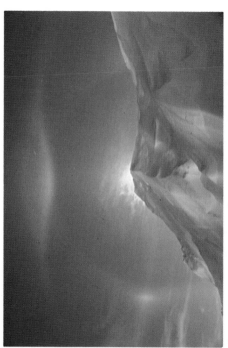

Plate 2-7. Display showing the upper tangent arc to the 22-degree halo for a sun elevation of 2 degrees. (Photographed in Wisconsin by A. James Mallmann)

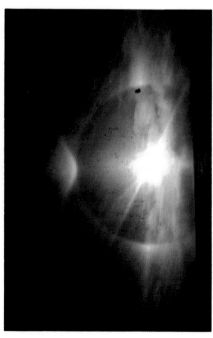

Plate 2-8. Upper tangent arc for sun elevation of 11 degrees. (Photographed in Antarctica by Vid Johnson)

Plate 2-9. Upper tangent arc for sun elevation of 23 degrees. (Photographed in Antarctica by Evan Noveroske)

Plate 2-12. Lower tangent arc for sun elevation of 17 degrees. (Photographed over Colorado by Alistair B. Fraser)

Plate 2-11. Lower tangent arc to the 22-degree halo (and subsun) for sun elevation of 11 degrees. (Photographed over Idaho by Alistair B. Fraser)

Plate 2-10. Upper tangent arc for sun elevation of 32 degrees. (Photographed in Wisconsin by the author)

Plate 2-14. Circumscribed halo for sun elevation of 48 degrees. (Photographed in Washington by Alistair B. Fraser)

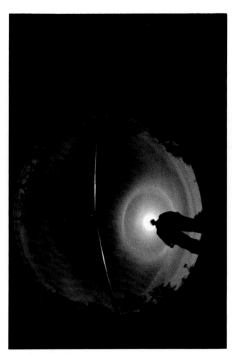

Plate 2-16. Upper tangent arc with upper suncave Parry arc showing as a distinct line; see Figure 2-16B. (Photographed in Arizona by Gary J. Thompson)

Plate 2-13. Upper and lower tangent arcs (circumscribed halo) for sun elevation of 40 degrees. (Photographed in New York by the author)

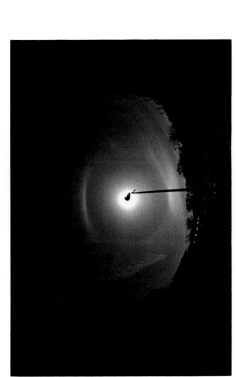

Plate 2-15. Circumscribed halo for sun elevation of 60 degrees. (Photographed in Wisconsin by the author)

Plate 2-19. Circumzenithal arc. (Photographed in Antarctica by the author)

Plate 2-17. Upper tangent arc with upper sun-cave Parry arc producing an area of approximately uniform intensity between the arcs; see Figure 2-16C. (Photographed in Antarctica by Bruce M. Morley)

Plate 2-20. Circumhorizontal arc. (Photographed in Colorado by Walt Tape)

Plate 2-18. Ice-crystal display. (Photographed in Antarctica by the author)

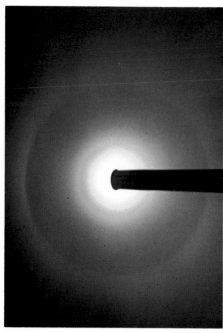

Plate 2-22. Eight-degree halo and 22-degree halo. (Photographed in Washington by Francis M. Turner)

Plate 2-23. Eighteen-degree halo and 22-degree halo. (Photographed in Pennsylvania by Gary J. Thompson)

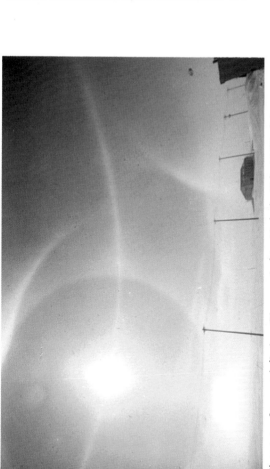

Plate 2-21. Ice-crystal display. (Photographed in Alaska by Takeshi Ohtake)

Plate 2-25. A halo puzzle, explained in the Appendix. (Photographed in Wisconsin by Walt Tape)

Plate 2-24. A halo puzzle, explained in the Appendix. (Photographed in Antarctica by Takeshi Ohtake)

Plate 3-1. Sun pillar. (Photographed in Wisconsin by the author)

Plate 3-2. Glitter path. (Photographed on the North Sea by the author)

Plate 3-3. Sun pillar and elongated sun dog. Simulation of both effects from the same set of flat-plate crystals. (Photographed in Wisconsin by the author)

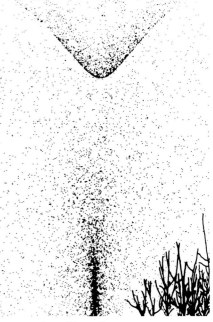

Plate 3-4. Sun pillar and upper tangent arc. Simulation of both effects from the same set of pencil crystals. (Photographed in Wisconsin by A. James Mallmann)

Plate 3-5. Light pillars over parking-lot lights. (Photographed in New York by Raymond F. Newell, Jr.)

Plate 3-6. The subsun composed of spots of light reflected from nearby ice crystals. (Photographed in Wisconsin by Walt Tape)

Plate 3-8. Twenty-two-degree halo, circumscribed halo, and parhelic circle for sun elevation of 50 degrees. (Photographed in Wisconsin by the author)

Plate 3-7. Subsun from an airplane. (Photographed by the author)

Plate 3-9. Subsun and 22-degree subparhelia (subsun dogs). (Photographed in Wisconsin by Walt Tape)

Plate 3-11. Anthelic pillar. (Photographed at Pt. Barrow, Alaska, by the author)

Plate 3-10. Anthelic arcs. (Photographed in Connecticut by Edgar Everhart)

Plate 3-12. Anthelic arcs. (Photographed in Alaska by the author)

Plate 3-13. Subanthelic (antisolar) arcs crossing the subparhelic circle. (Photographed in Washington by Lawrence Radkey, copyright 1972 by Alistair B. Fraser)

Plate 3-15. The subsun and Bottlinger's rings for a sun elevation of 12 degrees. (Photographed over Washington by Alistair B. Fraser)

Plate 3-14. A puzzle involving reflection. (Photographed by Francis M. Turner)

Plate 4-1. South Pole ice-crystal display. (Photographed by the author)

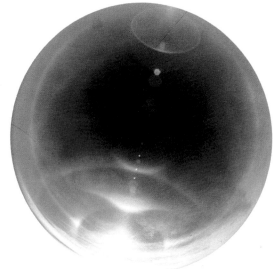

Plate 4-2. South Pole ice-crystal display. (Photographed by Austin W. Hogan)

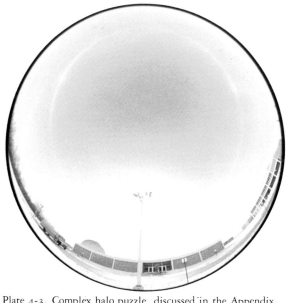

Plate 4-3. Complex halo puzzle, discussed in the Appendix. (Photographed in Illinois by Steve Bishop)

Plate 5-3. White clouds and black clouds. (Photographed in Ontario by the author)

Plate 5-1. Sky during the afternoon. (Photographed in Wisconsin by the author)

Plate 5-4. A series of clouds whose appearance is influenced by their background. (Photographed in Wisconsin by the author)

Plate 5-2. Night view of Plate 5-1. (Photographed by the author)

Plate 5-5. Yellow cloud. (Photographed in Ontario by the author)

Plate 5-6. Red clouds. (Photographed in Wisconsin by the author)

Plate 5-7. Blue clouds. (Photographed in Wisconsin by the author)

Plate 5-8. Air light in the Great Smoky Mountains. (Photographed by the author)

Plate 5-9. Crepuscular rays. (Photographed in Florida by Gerald Rassweiler)

Plate 5-10. Crepuscular rays above the sun. (Photographed in Wisconsin by the author)

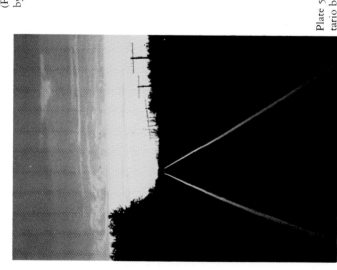

Plate 5-11. Railroad tracks showing the perspective of parallel lines. (Photographed in Ontario by the author)

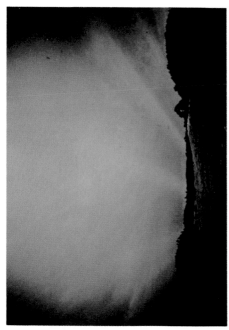

Plate 5-12. Sunlight through trees made visible by scattered light. (Photographed in Alaska by Takeshi Ohtake)

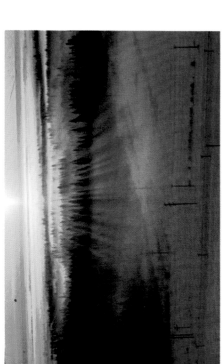

Plate 5-13. Anticrepuscular rays. (Photographed in Wyoming by Kenneth Sassen)

Plate 5-14. The twilight wedge. (Photographed in Ontario by the author)

Plate 5-15. The twilight wedge a few minutes later, showing the rising of the earth's shadow line. (Photographed by the author)

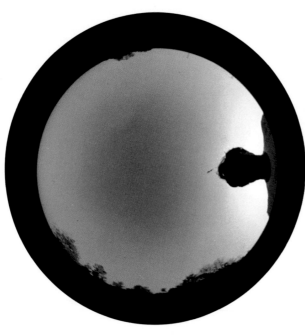

Plate 5-17. View of slightly hazy sky in Wisconsin. (Photographed by the author)

Plate 5-16. Fisheye view of clear sky at the South Pole. (Photographed by the author)

Plate 5-18. Scattering layer over the city of Paris. (Photographed by the author)

Plate 5-20. Light-scattering puzzle, explained in the Appendix. (Photographed in Wisconsin by the author)

Plate 5-19. Light-scattering puzzle, explained in the Appendix. (Photographed in Wisconsin by the author)

Plate 6-1. Corona aureole around the moon. (Photographed in New Zealand by the author)

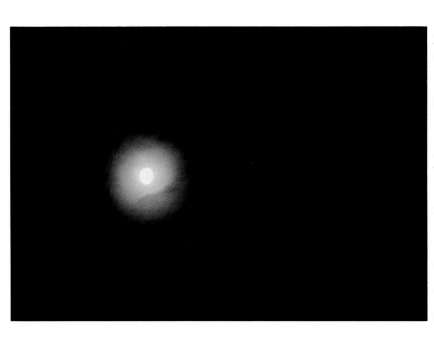

Plate 6-2. Corona around the sun. (Photographed in Antarctica by the author)

Plate 6-3. Corona around the sun showing well-defined orders. (Photographed by Gerald Rassweiler)

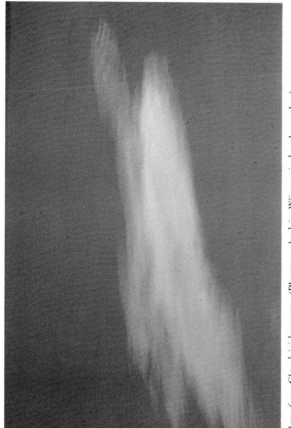

Plate 6-5. Cloud iridescence. (Photographed in Wisconsin by the author)

Plate 6-4. Roughness in corona circle resulting from variation of particle sizes in the cloud. (Photographed in Wisconsin by the author)

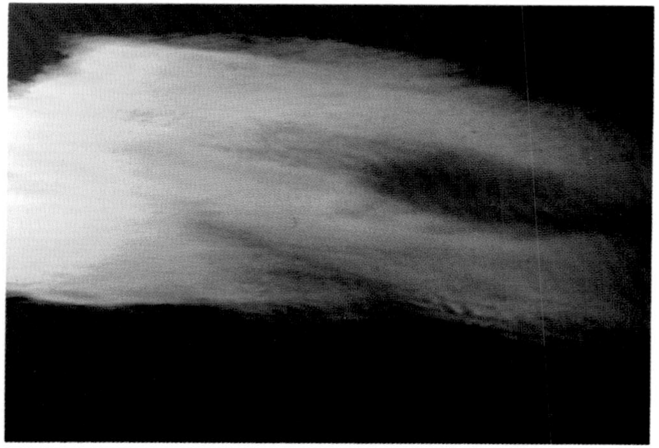

Plate 6-6. Cloud iridescence. (Photographed in Wisconsin by the author)

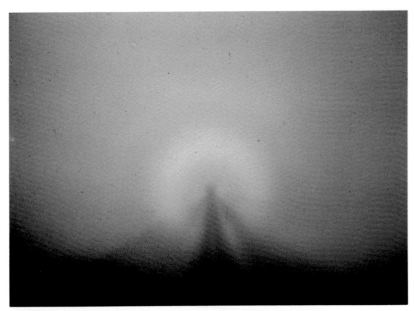

Plate 6-7. Specter of the Brocken. (Photographed in the Jura Mountains by June M. Gilby)

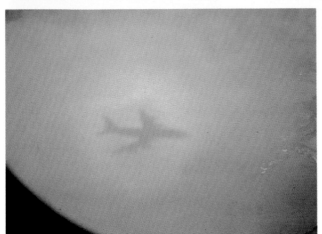

Plate 6-8. Glory around an airplane shadow. (Photographed over Ireland by the author)

Plate 6-9. Glory on distant clouds with the plane's shadow missing. (Photographed over the Atlantic by the author)

Plate 6-11. Dew heiligenschein. (Photographed by the author)

Plate 6-10. Noncircular glory at the edge of a cloud layer. (Photographed over the Atlantic by the author)

Plate 6-12. Heiligenschein streak. (Photographed in England by the author)

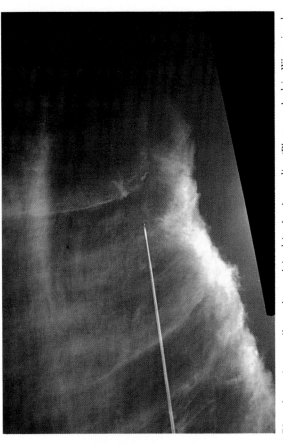

Plate 6-15. A contrail puzzle, explained in the Appendix. (Photographed in Wyoming by Kenneth Sassen)

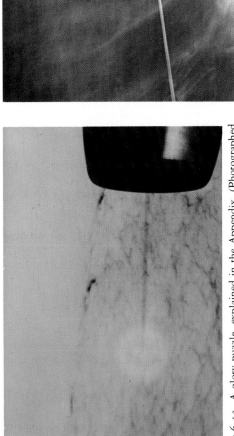

Plate 6-13. A glory puzzle, explained in the Appendix. (Photographed by Lewis Larmore)

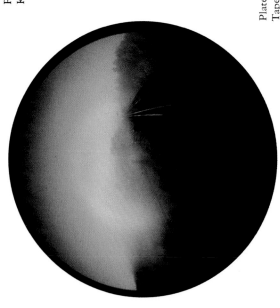

Plate 6-14. A glory puzzle, explained in the Appendix. (Photographed in Wisconsin by Walt Tape)

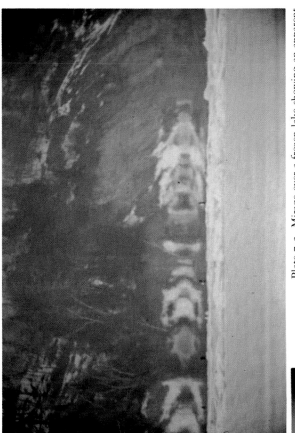

Plate 7-2. Mirage over a frozen lake showing an apparent reflection of features on the bluff. (Photographed in Wisconsin by the author)

Plate 7-3. Inferior mirage showing an apparent reflection of the distant shore. (Photographed in Wisconsin by the author)

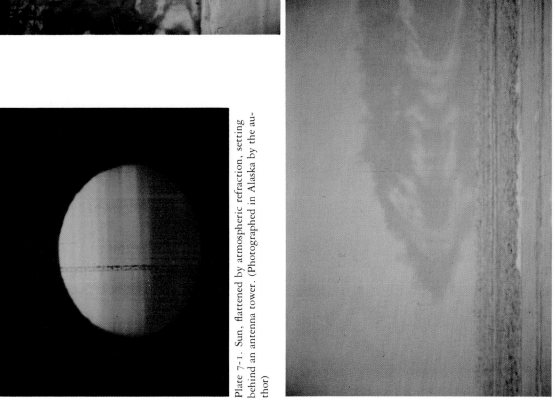

Plate 7-1. Sun, flattened by atmospheric refraction, setting behind an antenna tower. (Photographed in Alaska by the author)

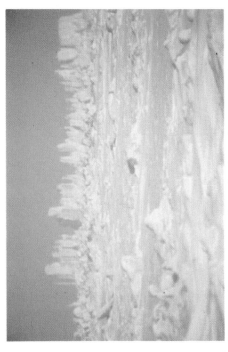

Plate 7-4. Fata morgana mirage showing an apparent wall over the Arctic Ocean ice. (Photographed at Pt. Barrow, Alaska, by the author)

Plate 7-5. Fata morgana mirage showing columns and spires over the Arctic Ocean. (Photographed by the author)

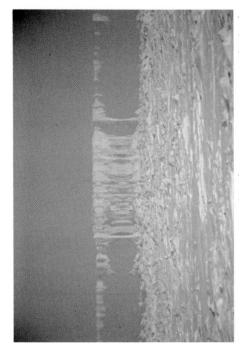

Plate 7-7. Another variation of the fata morgana mirage over the Arctic ice. (Photographed by the author)

Plate 7-6. Fata morgana mirage showing an apparent detached wall over the Arctic Ocean. (Photographed by the author)

Plate 7-8. Mirage of the Alaska Range. (Photographed by Glenn E. Shaw)

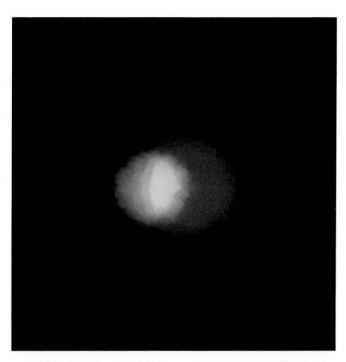

Plate 7-9. Enlarged photographs of Venus near the horizon: *left*, 10 degrees above the horizon; *right*, 2 to 3 degrees above the horizon. (Photographed in New York by Raymond F. Newell, Jr.)

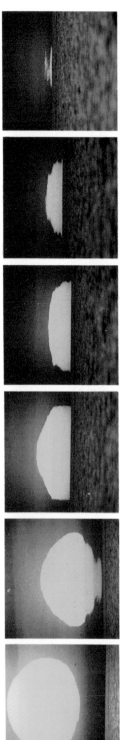

Plate 7-10. Sequence showing the distortion of the setting sun. (Photographed over Lake Ontario by Raymond F. Newell, Jr.)

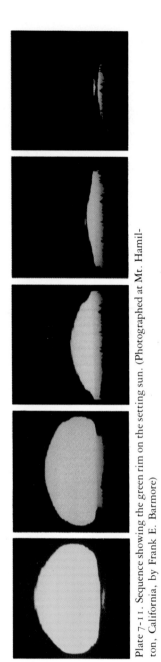

Plate 7-11. Sequence showing the green rim on the setting sun. (Photographed at Mt. Hamilton, California, by Frank E. Barmore)

2

Ice-crystal refraction effects: halos, arcs, and spots

We do not usually see rainbows in the wintertime, because at sub-freezing temperatures, angular ice crystals replace the spherical water drops that bring us rainbows. This trade-off only makes us richer, however. The interplay of sunlight with falling ice crystals produces a wealth of beautiful effects ranging from the common to the rare and from those that we understand to those that we still speculate upon. In this and the following two chapters, I attempt to organize some of our knowledge, and our ignorance, about these matters. These chapters on ice-crystal optical effects include more detail than some of the later chapters, largely because my colleagues and I have devoted considerable attention over the past several years to these effects, and, in some cases, we have contributed to their understanding. If there is more detail here than you are inclined to consider, please treat it indulgently and simply pick and choose.

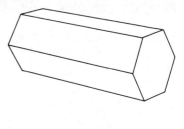

FORMS OF SIMPLE ICE CRYSTALS

The ice crystals most familiar to us are the complicated branching-crystal forms we commonly call snowflakes. Although snowflakes are most frequent near the ground in temperate climates, much simpler ice crystals can occur, with the appropriate combination of temperature and humidity, at other heights in the atmosphere. Figure 2-1 shows the simple hexagonal crystal forms that produce a surprising variety of optical effects in the sky and that will be the main focus of my attention for three chapters. Knight[1] is a place to start exploring the variety of ice-crystal forms and the conditions that lead to the growth of each.

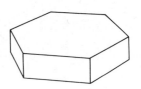

Figure 2-1. Two simple ice-crystal forms: *top*, a columnar or pencil crystal; *bottom*, a plate crystal.

A crystallographer would point out that the crystals of Figure 2-1 have the same crystal form; they are both hexagonal prisms and differ only in their length-to-width ratio. The difference, as we shall see, has profound consequences. I refer to the long, columnar crystal as a pencil crystal, after its resemblance to a common wooden pencil before sharpening. Let us call the other a plate crystal. Lest you

Figure 2-2. A magnification of small ice crystals collected as they fell from the sky. (Photographed by the author)

think that such idealized forms are the figment of some physicist's imagination, I show you Figure 2-2, a magnified photograph of some ice crystals trapped on a sticky plastic film as they fell. The largest hexagonal crystal in that picture is about 0.3 millimeter across, so that typical dimensions of many of those crystals are 0.1 or 0.05 millimeter. I collected these small "diamond dust" ice crystals at the South Pole research station in Antarctica, where they occurred right at the ground surface (or rather ice surface, which is about 3,000 meters above the ground surface). The same crystal forms do commonly occur in more temperate regions, higher in the atmosphere. Even in the summertime, the air temperature at moderate altitudes is below freezing, and we can see ice-crystal clouds displaying the halo effects I describe.

To understand what happens when ice crystals and sunlight interact, I need to organize and classify the different possibilities. In this chapter, I treat only those effects that arise when light passing through the ice crystals has its direction changed by refraction. (Chapter 3 deals with the effects resulting from reflection of the light by one or more flat crystal faces.) Within this chapter, I look first at a variety of effects that result from light passing through alternate side faces of the crystal and then at a different set of effects for light passing through a side face and an end face of the crystal (Figure 2-3).

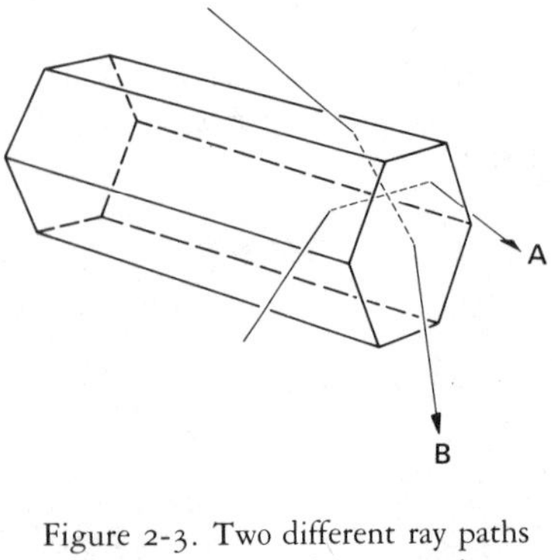

Figure 2-3. Two different ray paths through a hexagonal ice crystal.

In Figure 2-4, three of the side faces of a pencil crystal have been extended to illustrate that a light ray passing through alternate side faces is deviated exactly as if it had passed through a 60-degree prism made of ice. The magnitude of the deviation angle (shown as angle D in Figure 2-5) depends on the orientation of the ice crystal. If the crystal of Figure 2-5 were rotated clockwise, the deviation angle would increase. Surprisingly, if the crystal were rotated counterclockwise, the deviation angle would also increase. Figure 2-5, which shows the ray passing symmetrically through the prism, represents the case in which the deviation angle is a minimum; all other orientations increase this angle. For a 60-degree prism made of ice, this minimum angle is 22 degrees. Near the minimum-deviation orientation, the deviation of the light ray is insensitive to small rotations of the crystal; that is, the direction of the emerging ray changes very little as the crystal is rotated several degrees either way from the minimum position. As a consequence, when light rays pass

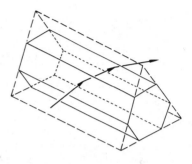

Figure 2-4. A light ray passing through a pencil crystal is refracted as if it were passing through a 60-degree prism.

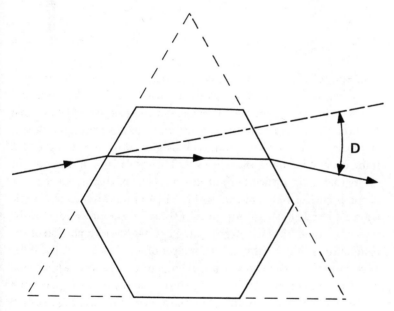

Figure 2-5. Refracted ray shows a minimum-deviation angle of 22 degrees.

through crystals with all possible orientations, there is a concentration of rays deviated by angles near 22 degrees. (This is the kind of argument used to explain the concentration of rainbow rays near the Descartes ray.)

To see a ray of sunlight deviated by 22 degrees directly to your waiting eye you would look at an angle 22 degrees away from the sun (Figure 2-6). The result of such refraction by a large number of ice crystals, tumbling through the air with random orientations, is a

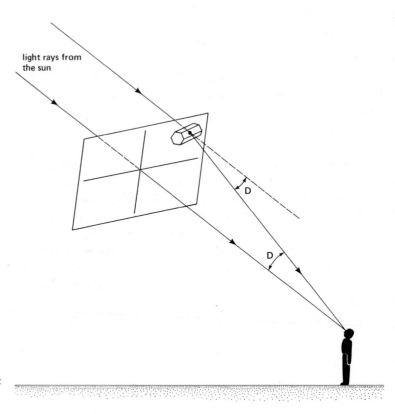

light rays from
the sun

D

D

Figure 2-6. To see sunlight that is
deviated by an angle D, you look at
an angle D away from the sun.

circle around the sun with the angular radius of the inner edge equal
to 22 degrees. Plate 2-1 shows such a halo — the 22-degree halo.

You will notice that the inner edge of the halo is reddish in that
picture. Remember that when white light passes through a prism,
all of the rays are deviated from their original direction, but the blue
light is deviated more than the red. The result is to spread out the
colors that make up white light into a spectrum that is, for me, one
of the beautiful sights of the world. The minimum-deviation angle
for red light is less than for green or blue; so the inner edge of the
red halo should be closer to the sun. Plate 2-2 shows a portion of the
halo, surprisingly bright, in a thin layer of ice clouds. The red inner
edge merges into a yellow band, then goes to a general white band
where all the colors overlap: This seems to be the normal appearance
of the halo. Do not be confused by the apparent difference in sizes of
the halos in Plates 2-1 and 2-2. The 22-degree halo is always the
same angular size, and its differing size in photographs results from
the use of different lenses in taking the photos and different degrees
of enlargement of the printed pictures.

TWENTY-TWO-DEGREE PARHELIA (SUN DOGS)

To get the complete halo we need to have all orientations of ice crystals present in the sky, so that at each part of the halo there are some

26

crystals present with the correct orientation to refract light to the eye. However, the ice crystals do not always fall with random orientations. The flat-plate crystal shown in Figure 2-1 will tend to fall with the axis connecting its flat bases vertical, the flat bases oriented nearly horizontally (if this seems strange, think of the way a leaf or a spread-eagled sky diver drifts to earth). Given a skyful of ice prisms with horizontal bases, we would not expect to see the entire halo. With the sun low in the sky, these plate crystals would have the proper orientation to refract light to the observer from the sides of the halo, but not from the top of the halo. The effect is shown in Plate 2-3. The better the orientation of the ice crystals, the smaller and brighter are the resulting spots on either side of the sun. In some cases they can be quite bright and of an apparent size comparable to that of the sun. Folk sayings have several names for these bright spots: false suns, mock suns, or sun dogs. Scientific folk sometimes call them parhelia. In Plate 2-4 you can see a bright sun dog to the left of the sun, with a tail of white light extending away from the bright, colored spot. The tail is the result of the rays that pass through the ice prism at angles other than minimum deviation; all the colors overlap, giving a streak of white light.

If you examine closely the magnified sun dog of Plate 2-5, you can distinguish the red, orange, yellow, green, and blue parts of the spectrum, beautifully exhibited in that little cirrus cloud. The reason that we can see a more complete spectral display in a sun dog than in the rest of the halo is a subtle one. I will sketch the argument for those of you who are motivated to think about it. The simplest treatment of light passing through a prism considers rays moving in a plane that is perpendicular to the refracting edge; in the case of our pencil crystals, this is the plane perpendicular to the long crystal axis (the axis corresponding to the lead in the pencil). Call that the normal plane, and call the rays that travel through the prism at some angle to the normal plane skew rays. For skew rays the angle of minimum deviation is greater than for normal rays, and the angle increases as the skewness increases.

This effect is manifested in the appearance of sun dogs for different elevations of the sun. If falling plate crystals are oriented with their bases horizontal, light from the sun on the horizon will pass through their normal plane and produce sun dogs with an inner edge 22 degrees away from the sun. But as the sun's elevation increases, the rays pass through these oriented plate crystals with more and more skewness, and the resulting sun dogs appear farther and farther from the sun. The effect is large enough that the sun dogs can easily be seen to lie outside the 22-degree halo for moderate sun elevation (Plates 2-9, 2-13, and 2-17). For a particular sun elevation, however, all of the oriented crystals have a definite minimum-deviation angle for each color, and they can produce a display like that of Plate

2-5. But consider what happens at some location on the halo as a result of refraction through randomly oriented ice crystals. Light is contributed to the halo from rays of all degrees of skewness, each one contributing a spectral pattern like that of Plate 2-5, but each pattern is located at a different distance from the sun. Because of the superposition of all these displaced patterns, all the colors are washed out except for the red inner edge.

This discussion of the difference in colors between the 22-degree halo and the sun dogs has grown somewhat complicated. It may help to recall the simple explanation developed to explain the presence of these effects: Randomly oriented crystals produce the 22-degree halo; and falling plate crystals, with bases horizontal, produce the bright spots on either side of the sun. The models are simple and yield a good understanding of the effects.

Chapter 1 described a quick way of making angular measurements that is ideal for seeking out and identifying the parhelia or parts of the 22-degree halo. Remember that for most people, the angle subtended by the outstretched thumb and little finger, with the arm extended, is about 22 degrees. So, with your thumb tip against the sun, your little finger tip should tell you where to look.

UPPER TANGENT ARCS TO THE 22-DEGREE HALO

Sometimes, in addition to the sun dogs and the 22-degree halo, we see an intensification of light at the top of the halo. The effect is fairly common and frequently takes the shape of a separate arc, tangent to the halo at the top but having quite a different curvature. The appearance of the arc varies greatly, depending on the elevation of the sun above the horizon. It cannot result from plate crystals, whose orientations produce light at the sides of the halo, but pencil crystals are another matter. These long crystals, falling in still air, tend to become oriented with their long axes horizontal (if that surprises you, throw a blade of grass in the air and see how it falls). Such a crystal orientation should give a concentration of light at the top of the halo, but it turns out to be quite difficult to predict the shape of the light pattern.

You might ask: Instead of trying to predict the shape of those arcs, why not just look and see what their shapes are? If we want to understand a phenomenon such as this, it is indeed important that we make careful observations. The problem, however, is that only by picking a model (in this example, light passing through hexagonal ice crystals as they fall with long axes horizontal) and then seeing whether the predictions of that model agree with the observations can we judge whether we have come up with a good explanation.

The main work done previously on these arcs (called the upper tangent arcs to the 22-degree halo) was published by two German scientists, Pernter and Exner,[2] in the early 1900s. They were able to plot a line representing the inner edge of the arcs. The shape of the arcs thus determined agreed well with the shape of the observed arcs, and they concluded that falling crystals with long axes horizontal produced these arcs. Wegener,[3] in 1926, was able to indicate roughly the extent of the arcs away from the inner (minimum-deviation) edge. It is interesting that Pernter and Exner assumed that plate crystals, falling on edge, would produce this effect and that pencil crystals, falling end down, with long axes vertical, would produce the sun dogs. They assumed that the crystals oriented themselves to minimize the air resistance as they fall; actually, they orient to maximize the resistance. Pernter and Exner's mistake about the orientation did not invalidate their general conclusions; it just assigned the effects to the wrong crystals.

THE COMPUTER SIMULATION OF ICE-CRYSTAL PHENOMENA

Modern scientists have a very powerful tool to use in problems involving ice crystals — a tool not available to earlier investigators. My colleagues and I have applied the computer to such problems.

We can begin by writing down a set of equations that describes a light ray going through an ice crystal, as illustrated in Figure 2-4. The equations are written for an arbitrary orientation of the crystal and an arbitrary elevation of the sun above the horizon. The only physical principle needed to determine the direction of the ray after it leaves the crystal is the law of refraction (Snell's law), applied both when the ray enters and when it leaves the crystal. The resulting general expression is quite complicated, and were it not for the existence of modern computers, we would probably give up at this point. However, the computer can be given this general expression and instructed to repeat the same calculation one hundred thousand times for one hundred thousand different orientations of the crystal. Knowing the direction of the light ray leaving the crystal is equivalent to knowing where to look in the sky for light coming to your eye from a crystal with that particular orientation. A convenient way to think about the process is illustrated in Figure 2-6. Consider the crystal to be located on a plane oriented at right angles to the line between your eye and the sun. If we know the direction of the light leaving the crystal, we can locate that crystal on the plane so that the light comes to your eye. The position coordinates of that location can be fed to a plotter, which places a spot on a sheet of paper at that location. The process is repeated again and again, each time giving the crystal another of the orientations we expect to find among the crystals as they fall through the sky. The resulting spot diagram represents the intensity pattern in the sky that we should see from light passing through a cloud of ice crystals having the distribution of orientations we have chosen for our calculation.

Actually, to obtain a good picture of the effect we have to concern ourselves with another set of details. Different amounts of light will get through crystals with different orientations, and we must consider several factors that affect the light intensity. Some of the light is lost by reflection from the crystal both on entering and on exiting from the crystal. We can calculate these losses exactly, using the angle of incidence of light on the surface and the index of refraction of ice. The cross-sectional area of light beam incident on the first crystal face gives another relative intensity factor. This area depends on the angle that the light makes with the face. In general, however, not all of the light that comes in the entrance face strikes the exit face inside the crystal; so we calculate the fraction of incident light that does get to the exit face. All of the factors combine to give us a relative intensity factor for light coming through the crystal in each orientation.

Consider, now, how we can take account of the varying intensities of the rays when the presence of a ray is represented by a spot on our spot diagram. One possibility would be to vary the size of the spots according to their calculated intensities. For our purposes, another approach seemed to be more convenient: We can take account of relative intensities by using these calculated values to dicard an appropriate number of the points whose positions we have calculated. For example, if we calculate an intensity factor of 0.4, we play a game of chance to decide whether to plot that point. The dice are loaded in this game to give only four chances out of ten that the point will be plotted. The game of chance is actually played mathematically in the computer by comparing the intensity factor with a random number between 0 and 1. Only where the factor is greater than the random number does the point get plotted.

In principle, once we can write a computer program to trace a ray through the crystal, we can simulate any of the phenomena that result from light taking that path through the crystal by specifying the appropriate orientations. For example, to specify any orientation of a hexagonal-prism crystal, we need to designate three angles. Two angles (azimuth and altitude) describe the direction in space along which the crystal axis is pointing, and the third angle describes how much the crystal is rotated about that axis. We can have the computer pick these angles at random and, by applying appropriate weighting factors or limits, can describe a distribution of orientations.

SIMULATION OF THE 22-DEGREE HALO AND SUN DOGS

First let us check our computer-simulation procedure by seeing if we can simulate the 22-degree halo and the sun dogs, which we understand fairly well. If we specify a set of angles that describes a group of crystals randomly oriented in the sky, we get the refraction effect shown in Figure 2-7. There it is: a halo of angular radius 22 degrees, with a sharp inner edge and an intensity that falls off as you

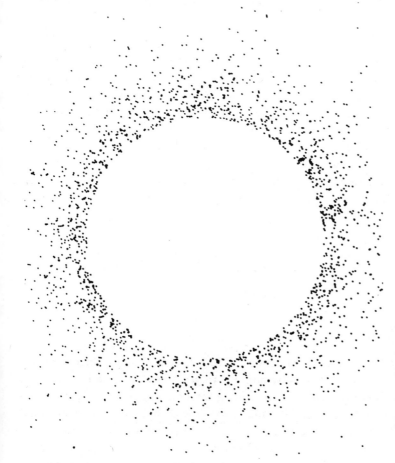

Figure 2-7. Simulation of 22-degree halo.

look away from that inner edge. It looks like a good start. In all of these calculations we are using the index of refraction appropriate to red light; so what we are really simulating is the form of the actual halo as photographed through a red filter.

To simulate the sun dogs we use plate crystals with their bases nearly horizontal, but allow all rotations of the crystals about the nearly vertical axes. The results are shown in Figure 2-8 for the sun at several different elevations above the horizon. In that figure, for

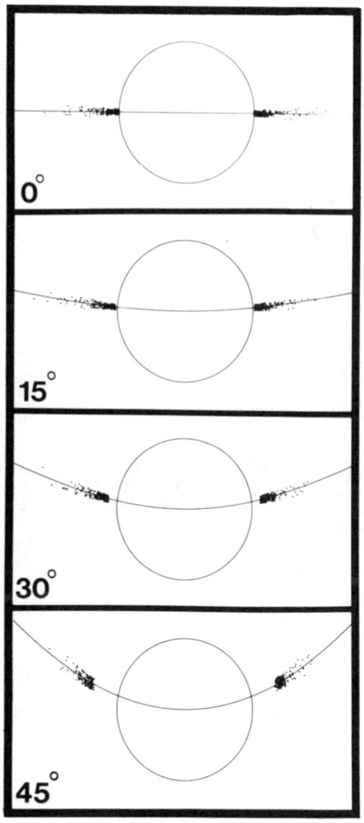

Figure 2-8. Simulation of sun dogs resulting from plate crystals with bases oriented horizontally (±3 degrees).

each simulation, a 22-degree circle is drawn to mark the position of the inner edge of the halo. We can see several interesting things from the simulations. As the elevation of the sun is increased, the dogs move away from the halo, as noted earlier. For high elevations of the sun the dogs take a parallelogram (or diamond) shape, which is, in fact, observed in nature. We find that the sun dogs disappear for sun elevation higher than about 60 degrees. One of the surprising effects in Figure 2-8 is the upward curvature of the sun dogs for all elevations of the sun above the horizon. This curvature appears in so many of the effects discussed later that it is worth understanding.

THE CURVATURE OF A LINE OF CONSTANT ELEVATION

The parhelia occur on each side of the sun, at the same elevation above the horizon as the sun. We might describe them as lying in a straight line, but they do not. A photograph shows this constant-elevation circle as a curve, as do the simulations of Figure 2-8. Think of it this way: A constant-elevation circle is like a large, horizontal hoop suspended over your head. When you look up at a section of the hoop, you see a curved line. If the hoop is lowered so that it is centered on your eyes, it would appear to be a straight line and would correspond to the horizon. Plate 2-6 is a multiple-exposure, wide-angle picture of the sun taken at the South Pole in January. At the Pole, the sun moves around the sky in a constant-elevation circle, giving 24 hours of sunlight each day during the summer half of the year. In this photograph the constant-elevation path of the sun is a curve, whereas the horizon is a straight line.

UPPER AND LOWER TANGENT ARCS AND THE CIRCUMSCRIBED HALO

The computer simulations of the halo and sun dogs, then, verify and augment our understanding of these effects. To explore the arcs that form as a result of light passing through alternate side faces of pencil crystals, as they fall through the air with long axes nearly horizontal, we started with the simpler case of keeping the long axis exactly horizontal. We allowed random rotation about the axis and random orientations of the axis in a horizontal plane (Figure 2-9). The results of the simulation, for different elevations of the sun, are shown in Figure 2-10 (the column labeled "Circumscribed halos"). Ignore for the moment the other columns; they are included in this figure because I will later want to compare them with the effect we are considering now.[4] In each case a 22-degree circle is drawn for reference and the horizon is represented by a heavy line. The perspective of each simulation is the same as that of a photograph centered on the sun. So this series of dot pictures embodies the consequences of the suggestion that some actual effects seen in the sky result from light passing

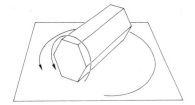

Figure 2-9. Distribution of pencil-crystal orientations that produce the circumscribed halo.

through pencil crystals with long axes horizontal. The test of the suggestion (or hypothesis, if you want so to dignify it) lies in the comparison of the simulations with arcs that actually appear in the sky. Plate 2-7 shows an upper arc for the sun near the horizon, comparable to the simulation for a sun elevation of about o degrees. As the sun increases in elevation to 30 degrees, the simulations show the *V*-shaped upper arc spreading and flattening with the ends drooping. Plates 2-8, 2-9, and 2-10 show this progression for sun elevations of 11 degrees, 23 degrees, and 32 degrees. At the sun elevation of 32 degrees, the upper arc flattens until at the top of the halo it curves neither up nor down but has a nearly horizontal center section. Plate 2-10 shows only a single arc, but it is not the 22-degree halo, which would be centered about the position of the sun just behind the tree top in the picture. Here is a case where all of the ice crystals are oriented so that we see no trace of the 22-degree halo, only the upper arc, whose shape is well matched by the simulation.

Note that the simulations show arcs tangent to the 22-degree circle both at the top and at the bottom. For low sun elevations the lower tangent arc lies completely below the horizon. Generally, when you look below the horizon, you are looking at the ground and will see nothing predicted by our simulations. You may be able to see these effects, however, in the situation where, looking down, you see a cloud of ice crystals — as when you fly in an airplane over an ice cloud. Plate 2-11 shows the lower tangent arc for a sun elevation of 11 degrees, photographed from such a vantage point. Note how the lower arc evolves in the simulations as the sun rises from 10 to 20 degrees above the horizon. Plate 2-12 shows the lower arc at a sun elevation of 17 degrees, in agreement with the simulation.

Plate 2-13 is a wide-angle photo with the sun at 40 degrees. The lower arc has flattened as predicted. Notice that in the 40-degree simulation the upper and lower arcs have curved around until they are joined on either side of the halo. At 40 degrees the intensity where these arcs join is low, and the complete curve is not seen in Plate 2-13. However, for greater sun elevations, the simulations predict and the photographs confirm (Plates 2-14 and 2-15) that the two arcs join to form a complete halo, completely enclosing the 22-degree halo and touching it at the top and bottom. Though it might be more appropriate to call this effect the circumscribing halo, tradition has given it the name circumscribed halo. Plates 2-14 and 2-15 show the circumscribed halo clearly separated from the 22-degree halo at the sides for sun elevations of 48 and 60 degrees. At higher elevations the circumscribed halo shrinks until it becomes indistinguishable from the halo it encircles. Note that the circumscribed halo and the upper and lower tangent arcs are really the same phenomenon, taking on different appearances for different elevations of the sun.

	CIRCUMSCRIBED HALOS	PARRY ARCS	ALTERNATE PARRY ARCS
0°			
5°			
10°			
15°			
20°			

34

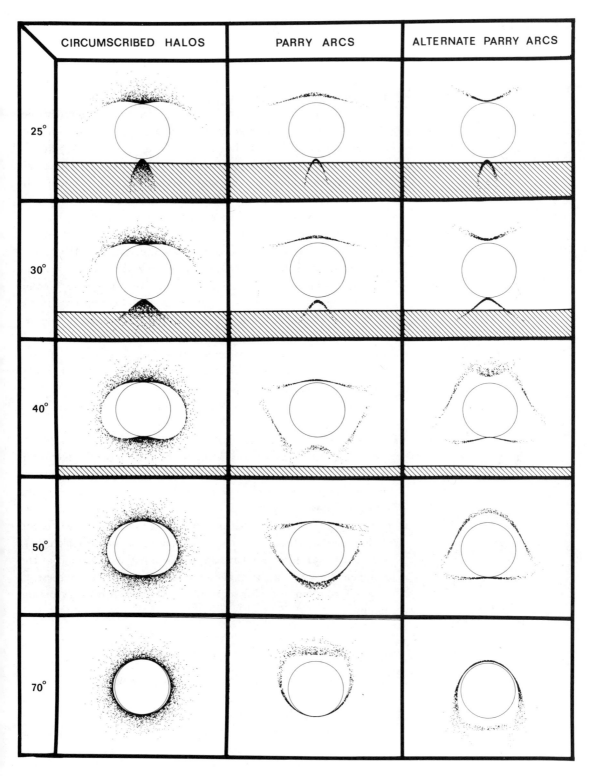

	CIRCUMSCRIBED HALOS	PARRY ARCS	ALTERNATE PARRY ARCS
25°			
30°			
40°			
50°			
70°			

Figure 2-10. Simulation of the circumscribed halo and Parry arcs

3 5

The photograph of Plate 2-15, incidentally, represents my home-made solution to a problem that vexes anyone who sees a wonderful sky effect and tries to photograph it with whatever camera happens to be available. Many of the effects cover such a large portion of the sky that a standard lens will capture only a disappointingly small portion of the whole. If your camera is a type that has interchangeable lenses, there are now some wide-angle lenses available that (for enough money) will solve the problem. You can go all the way to a fisheye lens (so called because of the similarity to the wide-angle view of the world obtained by a fish looking upward from below the water surface), which compresses an entire 180-degree field of view into a circle on your film. At the time I took the photograph for Plate 2-15, fisheye lenses were new and quite expensive and I didn't have one. I found that if I placed a reflecting sphere on the ground and looked into it from above, I could see, reflected there, the entire sky. By mounting my camera on a tripod over the sphere, I could photograph it. The tripod and camera are therefore reflected in the photograph (the camera shutter is open), and the entire 360-degree horizon is included in the circle. Actually, the reflecting sphere I used is not a very sophisticated optical element – it is an old Christmas tree ornament with a poinsettia painted on the other side. But one effect in a photograph taken with any fisheye lens is that the 22-degree halo is compressed vertically and so is not circular in the picture.

It seems to me that the photographs match the simulations impressively well; or rather – it is important to keep the distinction in mind – that the simulations match the pictures very well. The success of this simple model in predicting, in detail, the shapes of the upper and lower tangent arcs to the 22-degree halo and of the circumscribed halos for all the different sun elevations gives us some confidence that we have the correct explanations for these effects. It is clear that we need not have any tilts from the horizontal of the long pencil-crystal axes to explain the arcs. If small tilts are present, they will tend only to blur some of the effects shown in the photographs.

ORIENTATIONS OF FALLING ICE CRYSTALS

I have described three sets of phenomena that result from light passing through alternate side faces of hexagonal-prism ice crystals. Unoriented crystals produce the 22-degree halo; plate crystals produce sun dogs; and pencil crystals produce the tangent arcs and the circumscribed halo. Once you become convinced that plate crystals and pencil crystals do produce effects because of their high degree of orientation you might wonder how you can ever get the complete halo, which implies no preferred crystal orientation. Turbulent air, of course, could tumble crystals that otherwise would become orient-ed. My guess is that two other mechanisms are more important in disorienting crystals.

The aerodynamic forces that orient falling ice crystals depend on the size of the crystals. You can demonstrate that orientation by cutting hexagonal plates out of the low-density foam used for uphol-stery padding. Thin plates 2 or 3 centimeters across display the effect very well, but if you try to make them as large as 5 centimeters in diameter, they do not orient. I once wanted to make some crystals about 30 centimeters in diameter to demonstrate the effect, dra-matically, in front of a large audience, and I found that by lowering the density of the model crystals I could make them larger and still get the desired orientations as they fell. To get 30-centimeter mod-els that showed the effect, I had to construct hollow prisms out of thin sheets of polystyrene foam, which gave a very low overall den-sity. In a material with a density as high as that of ice, the size prism that best shows the orientation effects decreases to a fraction of a millimeter. So ice crystals in the approximate size range 0.05 to 2 millimeters assume special orientations while falling, whereas those that are much smaller, or larger, do not.

My polystyrene crystals also showed me that I did not get good orientations if the plate crystals were too thick for a given diameter. Perhaps crystals that are intermediate between plates and pencils, with length (thickness) about equal to diameter, can assume nearly random orientations as they fall through the sky. I will return to this point later in the chapter.

FREQUENCY OF HALO OBSERVATIONS

The three refraction effects I have discussed so far are fairly common, even though many people live their entire lives without noticing them. Over the past few years I have kept records which show that in the course of my usual daily rounds in Wisconsin, I see one of these effects on 70 or 80 days of a typical year. Most of the time I work inside, in an office or lab with no windows; so I am sure that these effects are also visible on many other days. (Contrast to this the three or four times a year when I see a natural rainbow.) The effects are also seen in hotter climates. On a hot summer day, whenever you look at wispy cirrus clouds, you are looking at ice crystals in an upper region of the atmosphere that is below freezing temperature.

Having described these three relatively common effects, I want to discuss two others that are fairly rare although they are produced by light taking the same path through the hexagonal crystals as those I have discussed. Both are named after the persons who reported their appearances with enough clarity and detail to attract the attention of later observers. They are called the Parry arcs and the arcs of Lowitz.

37

During his voyage in search of a northwest passage in 1819 and 1820, W. E. Parry[5] recorded his observations of flora, fauna, native people, geology, hydrology, and meteorology. He made sketches and gave detailed descriptions of a number of sun halo displays. Among other things, he described an arc above the sun that bears his name: the Parry arc. His original drawing is reproduced as Figure 2-11. The Parry arc is the arc labeled *c*, lying just above the upper tangent arc to the 22-degree halo. Subsequent sightings of the arc

Figure 2-11. Parry's original sketch. (From W. E. Parry. *Journal of a Voyage for the Discovery of a Northwest Passage*. Murray, London, 1821. Reprinted by Greenwood, New York, 1968)

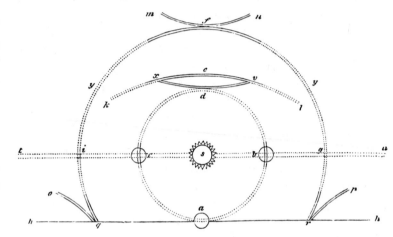

Figure 2-12. A and B: Two possible orientations of pencil crystals. C: A suggested ice-crystal form.

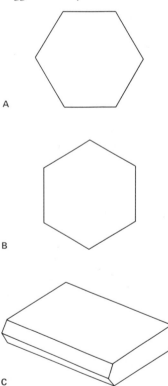

have not been very common, but they do occur and some photographs have been obtained. Plate 2-16 shows the 22-degree halo with an upper tangent arc, along with the additional arc that looks very similar to the effect in Parry's drawing. We have expanded on Parry's original observation and now suggest that his name be attached to arcs occurring both above and below the sun. Because these arcs occur near the upper and lower tangent arcs to the 22-degree halo, the two types are often confused; the distinction between them is discussed in a later section of this chapter.

Parry arcs are attributed to sunlight passing through the alternate side faces of pencil crystals whose orientations are even more restricted than those producing the tangent arcs or circumscribed halos. Not only is the long crystal axis confined to a horizontal plane, but rotation around the axis is restricted so that a pair of the faces (top and bottom) remains nearly horizontal (Figure 2-12A) or, it has been suggested, so that a pair of the faces (side faces) remains nearly vertical (Figure 2-12B). The aerodynamic forces for either of these orientations would seem to be quite small, and there is some disagreement about which orientation would be most stable. Ticker[6] suggests that an ice crystal of the form shown in Figure 2-12C would give the desired orientation effect, but I know of no

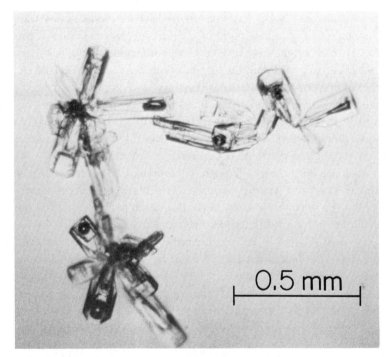

Figure 2-13. Clusters of crystals that might produce the Parry-arc orientations. (Photographed by Takeshi Ohtake)

direct evidence for the existence of that crystal form. It is possible that the crystals responsible for the Parry arcs are the hexagonal arms of a snowflake-type crystal, where the plane of the entire flake determines the orientation. Such a flake, in the appropriate size range, would tend to fall with its plane horizontal. Or it may be that the crystal is part of a cluster of crystals growing from a small seed crystal, as illustrated in Figure 2-13. The resulting orientation would be determined by the geometry of the cluster. I think that our computer simulations show what the appropriate orientations are; just how the crystals achieve the appropriate orientations is still an unanswered question.

The crystal distributions that I suggest as models to produce the Parry arcs are a restricted subset of the circumscribed-halo distribution. That means that all of the Parry-arc crystal orientations could be found in the circumscribed-halo group of crystals. Parry arcs must occur in the sky within the region of light intensity that is the circumscribed halo (or tangent arcs). Therefore, in order to separate the Parry arcs from these other effects, we must carefully compare, to see the differences caused by the more restricted Parry arc orientations.

Figure 2-10 shows, along with the circumscribed-halo simulation for various sun elevations, two other sets of simulations labeled "Parry arcs" and "Alternate Parry arcs."[7] The Parry-arc simulations come from pencil crystals with the rotational orientation suggested by Figure 2-12A, in which two of the side faces are maintained

39

nearly horizontally (for those results we have let the faces vary from the horizontal by a maximum of 3 degrees). The prediction of this model matches the descriptions and photographs that I have seen and so appears to explain the origins of the Parry arcs. Nevertheless, the column entitled "Alternate Parry arcs" in Figure 2-10 shows the predictions of the model using crystals with the orientations described by Figure 2-12B, in which a pair of side faces is held in a vertical orientation (to ± 3 degrees). By showing both of these series I do not suggest that both occur in nature. With both available for comparison with observations and photographs, however, we will be able to determine whether both sets of orientations do occur. Nonetheless, I will confine the rest of the discussion to the second column in Figure 2-10, the effects identified as Parry arcs.

Actually, there can be two distinct Parry arcs above the sun and two below, although not all will occur for all elevations of the sun. The origins of the two upper arcs are illustrated in Figure 2-14A by rays entering different faces of the oriented crystal. The ray labeled type 1 should give rise to the arc seen for higher elevations of the sun, whereas the type 2 rays should produce an arc for low sun elevations. A check of the simulations shows that only for sun elevations of about 10 or 15 degrees should we expect to be able to see both at once. Similarly, Figure 2-14B shows ways of producing two lower arcs, which should occur at different sun elevations. The simulation shows that both occur together only for sun elevations in the range of 40 to 50 degrees.

NOMENCLATURE OF THE PARRY ARCS

Most of the optical effects in the sky are named after their appearances — for example, circumscribed halo, upper tangent arc, rainbow, and circumzenithal arc. The effects named after people, such as the Parry arcs and arcs of Lowitz, are the exceptions to this system. Because the name Parry arc is generally accepted in the scientific literature, we should probably keep it, but I think that we can improve on the names of different kinds of Parry arcs. Goldie[8] has called them type 1, type 2, type 3, and type 4, as suggested in Figure 2-14A and B. They have also been identified (by Putnins[9]) by arbitrarily numbering the crystal faces (as in Figure 2-14C) and then giving two numbers to identify the entrance and exit faces. By this system the type 1 arc could be identified as the 1-3 arc. The difficulty is that these designations suggest neither the appearance of the arc nor the mechanism that produces it; I must return to Figures 2-10 and 2-14 to get the appropriate label whenever I want to refer to one of the arcs.

From the simulation of the arcs in Figure 2-10 we can see that the type 1 arcs are always concave toward the sun and the type 2 arcs are

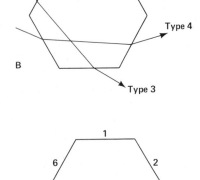

Figure 2-14. A and B: Rays that produce the four Parry arcs. C: Numbered faces for Putnins's designation of the Parry arcs.

always convex toward the sun. A similar distinction can be made for the type 3 and type 4 arcs. My colleagues and I have proposed to identify the arcs by their position above or below the sun and by their curvatures, calling them *suncave* arcs or *sunvex* arcs as they are either concave or convex toward the sun. For example, in Figure 2-10, the Parry arcs appearing for a sun elevation of 25 degrees would be called the upper suncave Parry arc and the lower sunvex Parry arc. The names are suggested directly by the shapes of the arcs. Figure 2-15 gives a description of the various arcs according to the three systems of nomenclature.

DISTINCTION BETWEEN PARRY ARCS AND TANGENT ARCS

Most of the Parry-arc pictures I have seen show the arc present along with the upper tangent arc. This superposition of the two effects may occur because the sunlight passes through two layers in the atmosphere, each containing crystals giving rise to one of the effects. The presence of the two effects superimposed in different proportions can give rise to different appearances. Figure 2-16 shows four simulations, varying from all upper suncave Parry arc (2-16A) to all upper tangent arc (2-16D). In 2-16B the Parry arc gives a distinct line above the tangent arc, as shown in Plate 2-16. In 2-16C the Parry arc seems to fill in the arc to give a well-defined region of approximately uniform brightness, as shown in Plate 2-17.

Some photographs of arcs identified as Parry arcs appear, by comparison with our simulation, to be upper or lower tangent arcs. Sometimes the distinction is clear, but at other times the identification can be tricky. Because the crystal orientations that lead to Parry arcs are a restricted set of the orientations that produce tangent arcs (or circumscribed halos), varying degrees of orientation can yield intermediate forms between Parry arcs and tangent arcs (or circumscribed halos). The Parry-arc simulations of Figure 2-10 result from crystals with maximum tilts of the top faces ±3 degrees from the horizontal. Figure 2-17 illustrates the effect of this maximum-tilt angle on the shape of the arc. As the maximum-tilt angle increases to ±30 degrees, the Parry arc turns into the circumscribed halo.

Sometimes it is difficult to distinguish between tangent arcs and Parry arcs without the intensity information provided by the simulations. For example, Figure 2-18B shows a brilliant arc below the sun observed by Jayaweera and Wendler[10] for the sun at an elevation of 25 degrees. There should be a difference in position between the upper edge of the lower tangent arc and the lower sunvex Parry arc of only about 1.5 degrees. On the basis of position measurement, the observers identified the arc as a Parry arc. Figure 2-18 includes

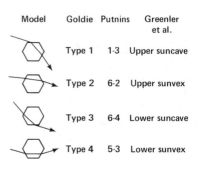

Model	Goldie	Putnins	Greenler et al.
	Type 1	1-3	Upper suncave
	Type 2	6-2	Upper sunvex
	Type 3	6-4	Lower suncave
	Type 4	5-3	Lower sunvex

Figure 2-15. Three different systems of nomenclature for the Parry arcs.

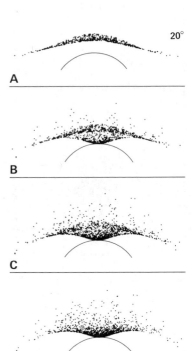

Figure 2-16. For a sun elevation of 20 degrees, A and D are an upper suncave Parry arc and an upper tangent arc; B and C are mixtures of the two effects.

41

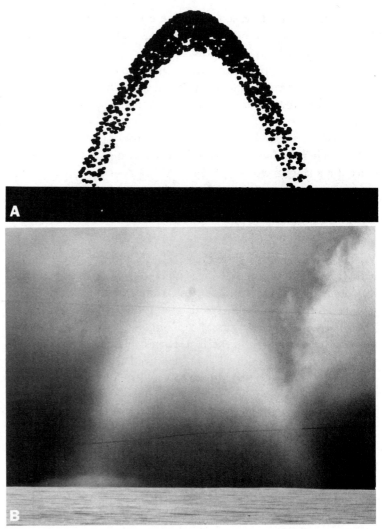

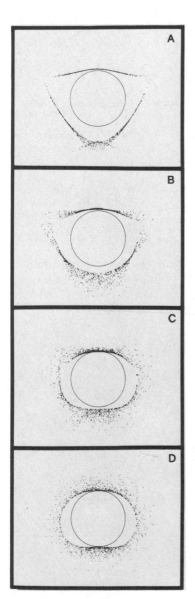

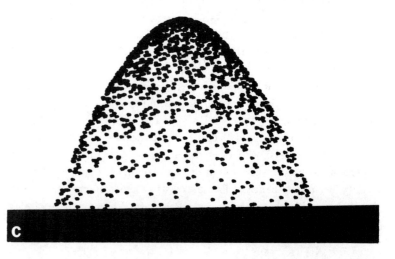

Figure 2-17. Simulations for a sun elevation of 45 degrees, showing the change from Parry arcs to the circumscribed halo as the orientations of the crystals become less pronounced. The long crystal axes are horizontal, but the top face tilts from the horizontal by up to (A) 1 degrees; (B) 6 degrees; (C) 16 degrees; (D) 30 degrees.

Figure 2-18. A: Simulation of Parry arc. B: Photograph showing an arc below the sun. C: Simulation of lower tangent arc. (B, from K. O. L. F. Jayaweera and G. Wendler. "Lower Parry Arc of the Sun." *Weather* 27, 50 [1972])

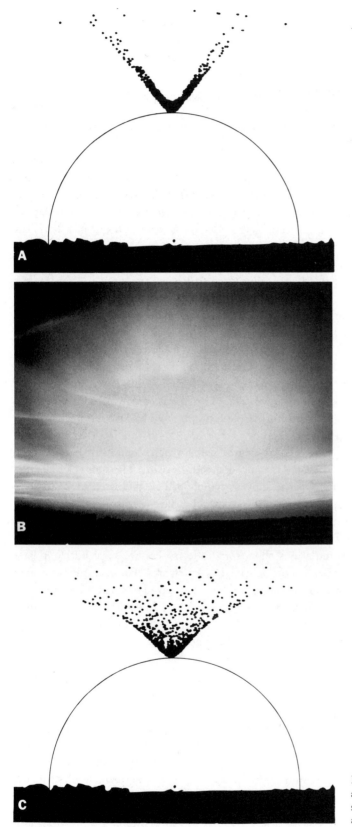

Figure 2-19. A: Simulation of Parry arc. B: Photograph of arc above the sun. C: Simulation of upper tangent arc. (B, photographed by the author)

two simulations done to the same scale as the photograph. Figure 2-18A is a simulation of a lower sunvex Parry arc produced by pencil crystals with the top face horizontal (± 5 degrees). The simulation of Figure 2-18C is of the lower tangent arc. It appears that the actual effect may be somewhere between these two extremes, but it is closer to the lower tangent arc; that is, there is, at most, a slight degree of ice-crystal orientation with respect to rotation around the horizontal axes.

A similar case (but one that leads to the opposite conclusion) is the photograph in Figure 2-19; the sun elevation here was determined to be 1.7 degrees. I assumed that the arc at the top of the halo in this picture was a tangent arc until we did Parry-arc simulations for this solar altitude. Figure 2-19A gives the result of a Parry-arc distribution with maximum tilt of ± 3 degrees, and 2-19C gives the tangent-arc simulation. The shapes and intensity distributions indicate that the arc in Figure 2-19B is a Parry arc.

Now that we know what Parry arcs look like, for all different sun elevations, I begin to think that they may not be as rare as we had supposed. For example, there is a trace of a Parry arc in Plate 2-9. As I look at the lower arc of Plate 2-11 and compare it with the simulations for a sun elevation of 10 degrees, I conclude that its intensity distribution agrees better with that of the Parry arc than with that of the lower tangent arc. The case is somewhat complicated by the superposition in that photograph of an additional bright spot of light right at the point of the arc. (You can see that it extends slightly inside the 22-degree halo, which should not happen with any light refracted by a 60-degree ice prism.) The spot is a reflection effect known as the subsun, which will be discussed in Chapter 3. These examples, and all the others I have seen so far, indicate that Parry arcs are produced by ice crystals with long axes horizontal and a pair of side faces nearly horizontal.

The simulations show some effects that we have not found reported, at least not identified as Parry arcs. For example, at a solar elevation of 40 degrees, the lower suncave arc divides into two arcs at the sides of the 22-degree halo. There is also the "wedding ring" effect for a sun elevation of 70 degrees. These simulations of as-yet-unobserved effects are predictions; the discoveries are yet to be made.

THE ARCS OF LOWITZ

Even less well understood than the Parry arcs are the arcs of Lowitz, named after Tobias Lowitz, who published a detailed description[11] of a magnificent, complex sky display seen in St. Petersburg, Russia, in 1790, and in the process gave it a permanent place in the records. No doubt countless other equally spectacular displays have passed from our memory for lack of a suitable chronicler. Lowitz's entire drawing is shown in Figure 4-2A in Chapter 4, but a portion of it is reproduced in Figure 2-24.

Lowitz described a pair of arcs that extended from the 22-degree halo, below the parhelic circle, up to the parhelia. Occasional observations have been reported, and a few photographs purport to show the Lowitz arcs. The most likely explanation of the Lowitz arcs seems to be the one advanced by Tricker,[12] which involves plate crystals spinning as they fall through the air.

It is not very convincing to explain an effect by introducing a strange crystal motion unless there is some physical basis for it, but actually it is easy to demonstrate a stable spinning mode for a flat falling object. If you hold a playing card or computer card horizontally, between thumb and finger, at the middle of a long side, and release it from that position by letting it roll off your fingertips, you will see it spin about its long axis as it falls. This rotation about the long horizontal axis of a flat object is a stable motion,[13] given objects of the right size and density. The suggestion[14] is that ice crystals, in the form of hexagonal plates, may spin about a major diagonal that remains horizontal as they fall through the air. Figure 2-20

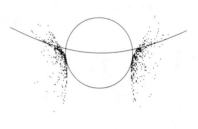

A

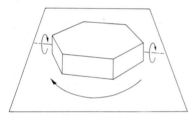

Figure 2-20. Distribution of crystal orientations suggested to explain the arcs of Lowitz.

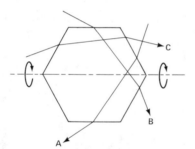

Figure 2-21. Three ray paths through a plate crystal spinning about a horizontal axis.

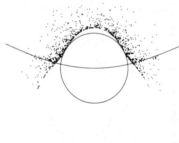

B

illustrates the orientation taken by such a crystal. There are three light paths that we should consider through alternate side faces of such spinning crystals, paths represented by rays A, B, and C in Figure 2-21. The simulations for each of these rays, passing through the spinning cyrstals, are shown in Figure 2-22 for a sun elevation of 30 degrees. Rays A give rise to arcs that seem to agree with Lowitz's description; rays B give rise to similar arcs that extend from the 22-degree halo above the parhelic circle down through the parhelia; and rays C produce an arc below the halo, nearly concentric with it. This description holds for sun elevations from about 10 to 40 degrees. At higher elevations arc C disappears.

According to this model, all of the arcs should appear together,

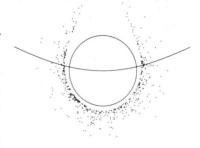

C

Figure 2-22. The three sets of Lowitz arcs resulting from the rays of Fig. 2-21.

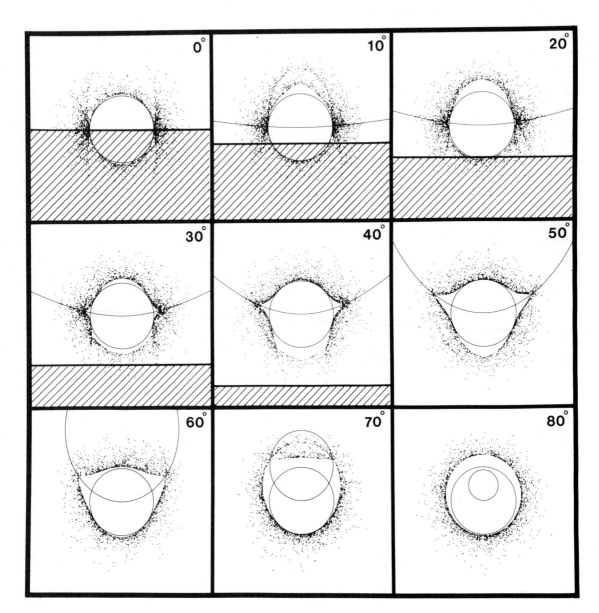

Figure 2-23. The simulations of Lowitz arcs from spinning plate crystals.

with the relative intensities shown. Figure 2-23 shows the combined simulations for nine elevations of the sun. You can see that the arcs extend outward through their intersection with the parhelic circle, an intersection that coincides with the inner edge of the parhelion. For low sun elevations the extensions through the parhelia may be the most obvious part of the display. My colleagues and I suggest that the arcs touching the 22-degree halo above and below the parhelic circle be called the upper and lower Lowitz arcs. Perhaps we need no name for the third component (Figure 2-22C), unless it is shown to be observable.

A serious objection to the theory I have just described is that

46

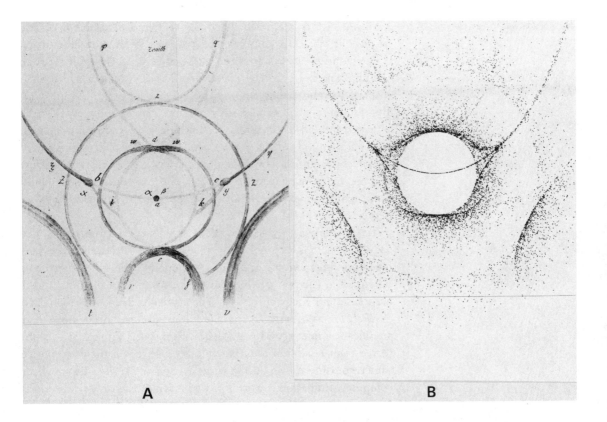

Lowitz reported only the lower arc, whereas our calculations indicate that the upper arc should be of comparable intensity. This is, indeed, a puzzle, but a composite simulation of several features seen in the St. Petersburg display may have given us the answer. Figure 2-24 shows a portion of Lowitz's drawing next to our simulation of several of the effects seen there. I will discuss the entire display in Chapter 4, but for now, look at the region of the Lowitz arcs where the 22-degree halo, the circumscribed halo, and the parhelic circle are all involved. For the sun elevation of 50 degrees you can see that the upper Lowitz arc is effectively obscured by its approximate coincidence with the circumscribed halo. Perhaps that coincidence explains its absence in Lowitz's drawing.

Figure 2-24. A: Part of Lowitz's drawing. B: Simulation of several effects in Lowitz's drawing.

This explanation for the mechanism of Lowitz arcs, though plausible, seems to me still tentative, awaiting some further photographic data for comparison. Figure 2-23 presents, in considerable detail, the predictions of the theory; now we need photographs for comparison.

THE 46-DEGREE HALO

This chapter has treated quite a number of refraction halos and arcs, all of which result from light passing through alternate side faces of hexagonal ice crystals. The variables that lead to different optical displays are the pencil or plate forms of the crystals and the resulting different orientations assumed by the crystals as they fall through the air. The basic optical effect for all these previously discussed displays is light passing through a 60-degree ice prism (Figure 2-4). But Figure 2-3 shows that in this same crystal light can also be refracted by a 90-degree prism. In both the plate and pencil crystals, the end faces and side faces intersect at 90 degrees, forming a prism for which the minimum angle of deviation is 46 degrees. The same argument that explained the 22-degree halo around the sun should also predict, or explain, another halo – that of angular radius 46 degrees.

Let me emphasize just what sky effect I am describing now. The 22-degree halo is a *big* halo (not to be confused with the smaller colored rings surrounding the moon – properly called the corona and described in Chapter 6). Its angular diameter is 44 degrees, and it occupies a large part of the sky. The 46-degree halo, however, is an *extremely big* halo, with an angular diameter of over 90 degrees. It is significantly rarer than the 22-degree halo, but I think that it is observed even less often than possible because it is difficult to notice something spread over such a large portion of the sky. To see if our computer simulation is working, we first trace rays of both the kinds

Figure 2-25. Simulation of 22-degree and 46-degree halos.

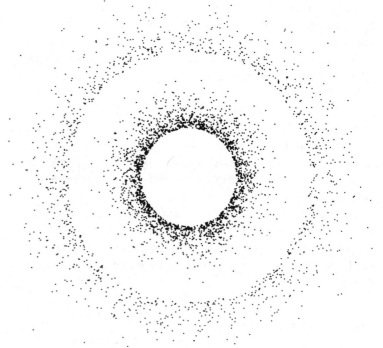

shown in Figure 2-3 for pencil crystals with random orientations. The results are the two halos of Figure 2-25, as predicted (though it should be noted that this casual "as predicted" does not mean that this, or any computer program of more than trivial length, can be expected to work correctly on the first try).

RELATIVE INTENSITIES OF THE 22-DEGREE AND 46-DEGREE HALOS

The relative intensities of the two halos are not correctly represented in Figure 2-25. The 46-degree halo is usually much fainter than the 22-degree halo for several reasons:

1. Only rays in a narrow angular range will get through the 90-degree prism (Figure 2-26A).
2. The transmitted rays for the 90-degree prism make small angles with the entrance and exit faces of the crystal, so that only a beam of small cross section will get through the crystal (Figure 2-26B).
3. Some of the light is lost by reflection at both entrance and exit faces of the crystal. Rays that make small angles with the crystal faces (have large angles of incidence) lose more intensity to reflection than do other rays; thus the reflection losses for the 90-degree refraction are greater.
4. The light producing the 46-degree halo is spread over a larger part of the sky than is that giving the 22-degree halo, and the effect is accordingly fainter.

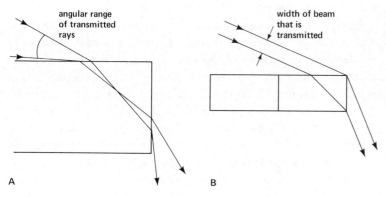

Figure 2-26. A: Narrow angular range of rays that can be transmitted by a 90-degree ice prism. B: The exit face of a plate crystal restricts the cross section of the transmitted beam.

These factors may explain why the 46-degree halo is seen considerably less often than its smaller counterpart, but there are probably other reasons why the 46-degree halo is not seen even at times when the 22-degree halo is brilliant. Sometimes pencil crystals have imperfect ends, so that they are adequate for the smaller halo but

49

deficient for the larger. I have heard one report of the large halo seen with no trace of the smaller one visible: I will leave the explanation of that observation to your speculation (several possible explanations occur to me, but I have no evidence to support any of them).

THE CIRCUMZENITHAL ARC

Each of the special set of crystal orientations that produces an optical display with the 60-degree refraction angle should also produce a unique feature with the 90-degree refraction angle. Once we have learned how to simulate the 46-degree halo we should be able to apply the same method to these other effects.[15]

We have seen that plate crystals floating with their bases horizontal produce sun dogs. What about rays that enter such crystals by an end face (base) and exit from the side, or vice versa, as shown in Figure 2-27? Suppose you hold a plate crystal in the sunshine and rotate it about the vertical axis connecting its flat bases. A ray that has entered from the top and exits from a side face changes its direction as the crystal is rotated, but in a very special way. The angle between the ray and a vertical line does not change; as the crystal rotates, the ray rotates around the vertical but keeps the same angle with the vertical. If we transfer our attention from the one crystal to the observer, and ask where he or she looks to see light from a skyful of such oriented crystals, the answer is straightforward: the observer looks in directions that make a constant angle with the vertical. The vertical direction in the sky is called the zenith; so that effect we are discussing should be an arc that lies on a circle centered on the zenith. You could hardly invent a more descriptive name for this effect than the one it bears: the circumzenithal arc.

Plate 2-18 shows a photograph that includes the circumzenithal arc. A striking feature of this arc frequently is the rainbowlike brilliance and saturation of the spectral colors. The spectral colors are much more vivid than those in either of the refraction halos. The halos are minimum-deviation effects; each color has a concentration at its minimum angle of deviation but has a significant amount of light refracted at greater angles, overlapping other colors in the spectrum. (I noted earlier how the halo colors are further diluted by skew rays, which have larger minimum-deviation angles than rays passing through a normal plane.) The circumzenithal arc, on the other hand, results from crystals with an additional degree of orientation and does not depend on a minimum-deviation effect for the concentration of each color.

If we simulate the circumzenithal arc, we get spot diagrams that are in good agreement with the observed effects. The simulations discussed previously (with the exception of Figures 2-16, 2-18, and 2-19) have all been done with the sun at the center; that is, the plane

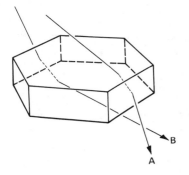

Figure 2-27. Light rays that produce (A) the circumzenithal arc and (B) the circumhorizontal arc.

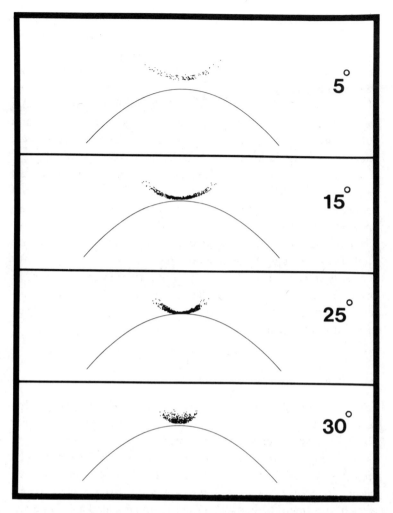

Figure 2-28. Simulations of the circumzenithal arc.

shown in Figure 2-6 is perpendicular to the observer – sun line. These simulations should match the perspective of a photograph taken with the sun in the middle of the picture. If the camera is pointed in a different direction, the perspective of the picture is changed; for example, the 22-degree halo will not record on film as a circle when the camera is pointed away from the sun. We can, however, shift the reference plane of Figure 2-6 to make it perpendicular to a line pointing in some other direction. The resulting plot of points will then reproduce the perspective of a photograph taken with the camera pointing in that direction.

The simulations of Figure 2-28 represent the circumzenithal arc for several elevations of the sun. They are plotted to match the perspective of photographs taken with the camera pointed at the inner edge of the 46-degree halo, which is represented by a line in the simulations. Remember that we are simulating only the red-light portion of the arc. The differing index of refraction for other colors gives similar arcs farther from the sun, spreading out the display into the complete spectrum. Plate 2-19 shows the arc for a sun elevation of 21 degrees with a perspective that should be matched by the simulations of Figure 2-28. At the time of this photograph, apparently all of the ice crystals were oriented plate crystals, which yielded no trace of the 46-degree halo. The arc comes near the 46-degree halo but is not really tangent to it, except at a sun elevation of 22 degrees, when the center portion of the arc is at the minimum-deviation distance from the sun. Because of the proximity of the arc to the 46-degree halo it is sometimes misidentified as a contact arc or a tangent arc to that halo. For a sun elevation of 15 degrees, the simulation shows that the arc extends slightly less than a third of the way around the zenith (108 degrees of azimuth). The circumzenithal arc can be produced only when the sun elevation angle lies in the range illustrated in Figure 2-26a; that is, it can appear only when the sun is less than 32 degrees above the horizon. In my experience the circumzenithal arc is the most commonly observed 90-degree refraction effect; I see it more often than any part of the 46-degree halo.

THE CIRCUMHORIZONTAL ARC

Let us now consider the other ray shown in Figure 2-27, the one that comes in a side face and out the bottom face of the plate crystal. For all rotations of the crystal about a vertical axis, this ray emerges at the same angle with respect to a horizontal plane; hence it should be seen at a fixed elevation above the horizon independent of crystal rotation. It produces an arc of light that lies at a constant elevation, stretching around the horizon – the circumhorizontal arc. Figure 2-29 shows the simulations for several sun elevations. Like the circumzenithal arc it lies near the 46-degree halo but touches it only for one sun elevation, 68 degrees. It can be seen only when the sun is high in the sky, at an elevation greater than 58 degrees above the horizon. It is therefore seen more commonly in lower than higher geographical latitudes, although it can still be observed in the summertime in the northern United States, southern Europe, and Japan. It has many of the properties of the circumzenithal arc, and its bright spectral colors can make it a spectacular display, extending almost a third of the way around the horizon (Plate 2-20).

SUPRALATERAL AND INFRALATERAL ARCS

The simulations of the circumzenithal and circumhorizontal arcs hold no particular surprises because they can be fairly well understood by other, simple calculations. The rays that go through the

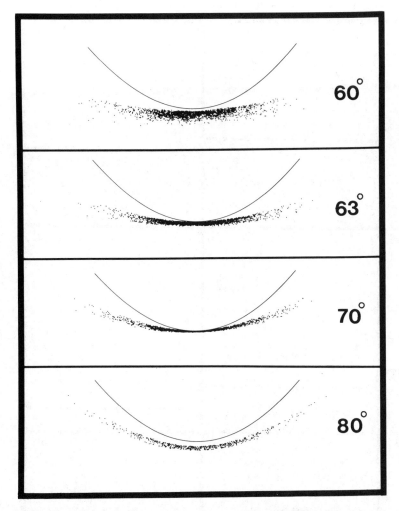

60°

63°

70°

80°

Figure 2-29. Simulations of the circumhorizontal arc.

90-degree prism faces of the pencil crystals as shown in Figure 2-30 are another matter. If the crystals are held with long axes horizontal, but allowed all rotation about these axes, we have the same distribution that accounts for the circumscribed halo. The two rays of Figure 2-30 would contribute to two arcs, one generally above the sun and the other below, but the exact shape and intensity distribution of these arcs are difficult to determine by the usual calculations. Both of the arcs should lie outside the 46-degree halo and would be expected to touch it for crystal orientations that result in minimum deviation through the 90-degree prism. The upper arc is commonly called the supralateral arc; and the one located mainly below the sun, the infralateral arc. Because the same crystal, with the same well-formed faces, is required to form either arc, we would expect to see both at the same time, and both are shown together in Figure 2-31. A 46-degree-radius circle and the parhelic circle are added for reference in these simulations. (The parhelic circle is the circle of constant elevation above the horizon passing through the sun.)

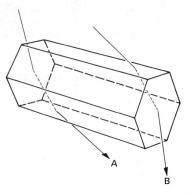

Figure 2-30. Light rays that produce (A) the infralateral arc and (B) the supralateral arc.

53

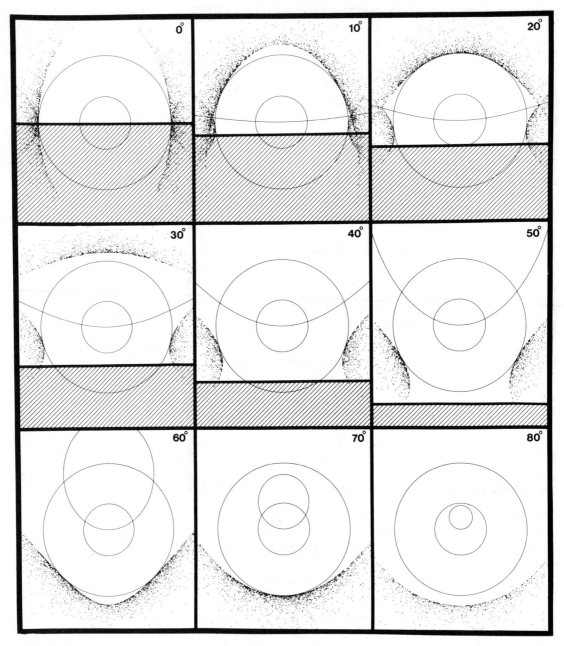

Figure 2-31. Simulations of the supralateral and infralateral arcs.

The supralateral arc disappears for sun elevations greater than 32 degrees (a matter understandable by reference to Figure 2-26). It does lie mostly above the sun for sun elevations greater than about 15 degrees, but at lower elevations it divides into two arcs, one on either side of the sun. Similarly, the infralateral arc lies largely below the sun for elevations around 90 degrees, but it splits into arcs on either side of the sun at lower elevations. For low sun elevations in the simulations of Figure 2-31, the supralateral and infralateral arcs

54

overlap; but if you trace the evolution of the arcs from one sun elevation to the next, you can easily sort out the two effects.

Is there any physical reality to all this? A number of visual sightings of the supralateral arc have been discussed, but good photographs are scarce. The increasing availability of very wide-angle lenses should improve the chances of photographing this arc. Several photos of the infralateral arc have been made, giving us observational data to compare with our simulations. Most of these seem to be of the divided arc in the 30- to 50-degree range. I have seen neither predictions nor descriptions that look like the low sun elevation simulations.

Figure 2-32 shows a photograph containing an arc that appears to be an infralateral arc. From the time of day and the photographer's location,[16] we calculate the sun's elevation to be 38 degrees. Knowing the focal length of the lens (35 millimeters) used to take the picture, we can measure angles on the photograph. We estimate that the center of the picture is located about 6 degrees below the parhelic circle and at an azimuth of 48 degrees to the right of the sun (note that 48 degrees is the azimuthal angle, measured along the ho-

Figure 2-32. A: Photograph of circumscribed halo, infralateral arc, and parhelic circle. B: Matching simulation. (A, photographed by Edgar Everhart)

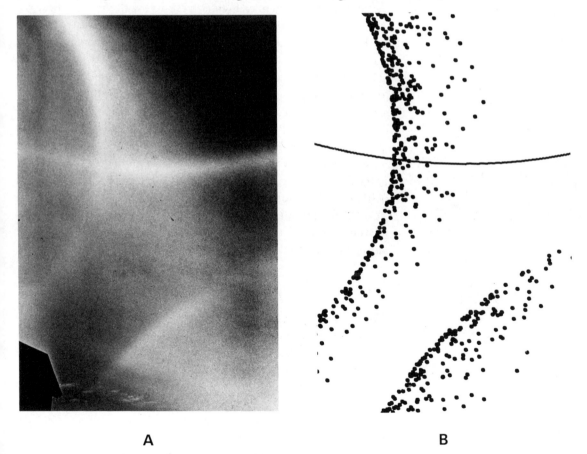

A   B

55

rizon, not the angular separation, measured along a great circle). Figure 2-32 also shows a simulation made to match the perspective of this picture, which includes the infralateral arc, the circumscribed halo, and a line representing the parhelic circle. A reflection effect, to be discussed in Chapter 3, gives a colorless arc along the parhelic circle that is visible in the photograph. This simulation matches the form and intensity distribution of those effects to a surprising degree. Other pictures and drawings of infralateral arcs of this shape, for sun elevation in the range of 30 to 50 degrees, seem to be well represented by this model.

Plate 2-21 shows a remarkable arctic halo display including several of the features already discussed. For a sun elevation near 30 degrees, the infralateral arc shows brilliantly. On the original slide the supralateral arc is visible though probably too faint to survive reproduction.

I saw the photograph reproduced as Figure 2-33A shortly after we had done the simulations of the infralateral and supralateral arc, and my attention was drawn to a faint feature just to the right of the 46-degree halo arc. Without the simulations I probably would not have noticed the faint area of triangular intensity, with the base of the triangle lying on the 46-degree halo and the peak located on the parhelic circle to the right of the halo. Detecting this triangular feature, faint even in the original, may require a bit of imagination. The perspective of a wide-angle photograph does some funny things to familiar effects, but by picking the appropriate direction for the center of our simulation we should be able to reproduce the effects. The simulation of Figure 2-33 shows our attempt to match the photo with the 22-degree halo, the upper tangent arc (skewed to the right by the wide-angle prespective), a faint 46-degree halo, and the infralateral and supralateral arcs for the sun elevation of 23 degrees. Although the comparison is not beyond debate, it appears to me that the faint triangular feature is explained by the shape of the simulation. I suspect that in the future we will see some of the other forms predicted in Figure 2-31, but not yet recorded.

THE CRYSTALS THAT PRODUCE THE 46-DEGREE HALO

The simulations of Figure 2-31 show that for sun elevations between 10 and 20 degrees the supralateral arc comes rather close to matching the upper part of the 46-degree halo. Small tilts of the horizontal crystal axes would improve the fit to the extent that, seeing only the portion of the display above the horizon, you might have difficulty distinguishing between the supralateral arc and the 46-degree halo. Some scientists think that we seldom or never get the randomly tumbling crystals that would give us the complete 46-degree halo but see instead only portions of the halo produced by

Figure 2-33. A: South Pole photograph by Bruce Morley. B: Simulation matching several features of the photo.

fairly well-oriented pencil crystals. Fraser [17] argues that plate or pencil crystals would require diameters less than about 7 micrometers (7 millionths of a meter) to be randomly oriented (in the absence of turbulence), but for crystals this small, diffraction effects would dominate the interaction of light with the crystals and would prevent the formation of the large halo. He suggests that the complete 22-degree halo can be produced by the amount of tipping occurring for plate crystals in the 15- to 40-micrometer range but that we cannot get randomly enough oriented pencil crystals to get the complete 46-degree halo.

In support of this argument it is claimed that few people have actually seen the entire 46-degree halo. Of course, for the entire halo to be seen from the ground, the sun would have to be higher than 46 degrees above the horizon. My first reaction is that I have seen and have even photographed the 46-degree halo below the sun, but another look at Figure 2-31 shows that for sun elevations of 60 to 70 degrees, the infralateral arcs form a reasonable facsimile of the halo below the sun. Without a conscious check to see whether the entire halo is actually there, it is impossible to answer the question later; it is very easy in retrospect for the mind to sketch in a missing portion of the arc. Obviously what is needed are good observations, preferably backed up with wide-angle photographs. My own guess is that randomly oriented crystals do produce the entire halo and that fairly large ice crystals, whose length and width are about equal, may contribute to the intensity. Such crystals, which are intermediate between the pencil and the plate forms of the hexagonal prism, would experience only small aerodynamical orientation forces. That is, of course, only speculation: The proof of this particular pudding would be found by examining the ice crystals present in the atmosphere when a complete 46-degree halo was present.

CONTACT ARCS TO THE 46-DEGREE HALO

We have previously investigated light going through the alternate side faces of spinning plate crystals to determine whether this model accounts for the arcs of Lowitz. If those spinning crystals do exist in significant populations in the sky, we would also expect to see a series of arcs resulting from refraction through the 90-degree faces of the crystals. There are six paths through the crystal that we must investigate, the paths represented by the rays shown in Figure 2-34. Those three pairs of rays are not equivalent; rays A, B, E, and F involve the side faces adjacent to the axis of rotation, whereas rays C and D involve the side face parallel to the axis. You would expect each of these rays to give rise to an arc that would, at some point, touch the 46-degree halo. Figure 2-35 shows the simulation results for each ray for a sun elevation of 40 degrees. I will refer to these arcs as the contact arcs to the 46-degree halo. Ray C has been considered in the past to explain arcs at the top of this halo. Such an arc is difficult to identify without knowing the shape of other arcs occurring in the same region of the sky, such as the circumzenithal arc and supralateral arc. With this spinning-crystal model, the crystals that give the effects of ray C should give the others at the same time. Our simulations also indicate that all of the contact arcs should be of comparable intensities; the same number of trial crystal orientations were used for the simulation of each arc.

Figure 2-36 shows the complicated set of effects that results when arcs are superimposed, for each sun elevation, covering the range from the horizon to 80 degrees. With the sun 30 to 40 degrees above the horizon you might not be able to distinguish the set of

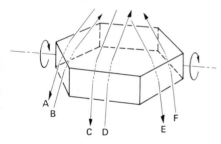

Figure 2-34. Rays that should produce contact arcs to the 46-degree halo for spinning plate crystals.

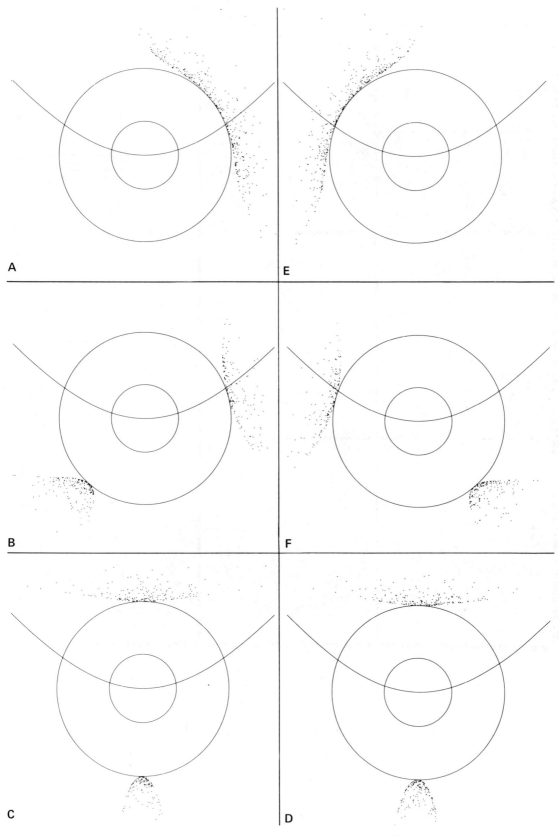

Figure 2-35. The contact arcs produced by the rays of Figure 2-34.

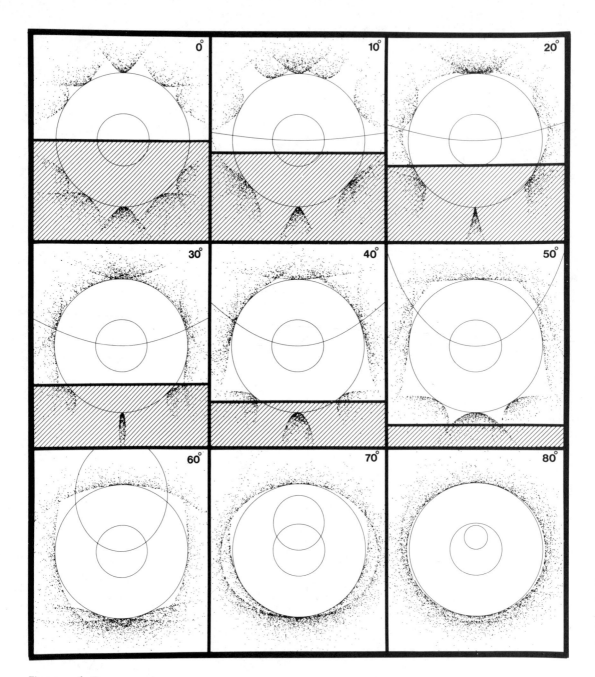

Figure 2-36. Contact arcs that would result from spinning plate crystals.

contact arcs from the 46-degree halo. The same problem might occur with very high sun elevations (80 degrees). I have never seen anything that looks like those simulations. On the other hand, I am very aware of the selective way in which we notice familiar things and miss obvious things that are not familiar. If well-developed spinning plate crystals exist in the sky in significant numbers, Figure 2-36 shows some of the effects they should produce. The search for the predicted effects is a way of determining whether those spinning crystals exist.

RARE ICE-CRYSTAL HALOS

The 22-degree halo is probably never noticed by a majority of people in the world today; it is seen occasionally by a sizable minority; and it is seen frequently by only a small minority. Probably only a small minority have ever seen the 46-degree halo, and still fewer have seen it several times. There are some other halos that are seen only rarely, even among the small group of hard-core sky watchers, and are seldom (or never) well photographed. We will consider some of these rare halos, but it is inevitable that, as the observations and photos become scarcer, the discussion becomes more speculative.

A scattering of reports in the scientific literature indicates a halo with angular radius near 90 degrees. It is called Hevel's halo, and suggestions for its origin involve combinations of reflections and refractions in ice crystals having the form of right hexagonal prisms (I will discuss it further in the next chapter). It seems unlikely, however, that such a crystal can account for the variety of other rare halos that have been reported, including a halo of radius 8 or 9 degrees, another of radius 17 or 18 degrees, some with radii slightly larger than 22 degrees, and one with a radius between 30 and 35 degrees. For some time I was skeptical of the existence of most of these halos. In some pictures the reported halo was lost (to my eye) in the general brightness near the sun, and in others, the halo looked suspiciously like a case of lens flare — an artifact in the picture resulting from multiple reflections within the camera lens system. (You can avoid some of these problems, should you be photographing a rare halo, by moving so that the sun is covered by a lamp post, chimney, telephone pole, flying goose, or other appropriate obstacle.) The photos of Plates 2-22 and 2-23, however, are convincing to me. I measure the radius of the smaller halo on Plate 2-22 as 8 degrees and that of the smaller halo on Plate 2-23 as 18 degrees.

Another set of evidence concerns an unusual halo display seen in England on Easter Sunday, April 14, 1974. Observations were collected from twelve independent observers, each of whom observed at least two concentric halos simultaneously, and as many as six halos were reported from some places. In several cases, the reports were backed up by photographs. These observations were gathered and analyzed by Goldie, Meaden, and White.[18] So, even to a skeptical reader, the evidence for the existence of several halos with radii between 8 and 35 degrees seems substantial. An article by Tricker[19] further discusses these rare halos and contains one of the color photographs of the Easter Sunday display.

61

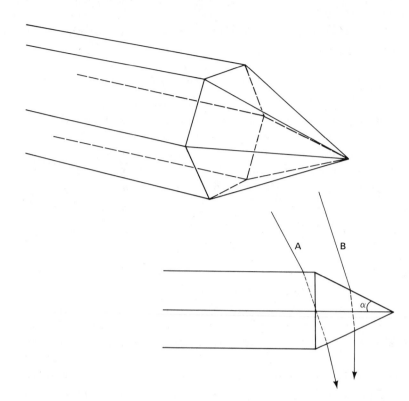

Figure 2-37. Hexagonal crystals with pyramidal end faces that may produce a variety of rare circular halos.

There seems to be no way to explain these halos with the right hexagonal ice prisms that have been the key to so many effects. The usual explanation for these unusual halos invokes hexagonal ice crystals with pyramidal end faces, as shown in Figure 2-37. Depending on the angle between the long axis and the end faces (α in Figure 2-37), we could explain almost any size halo. For example, if α were 25 degrees, rays taking path A through the crystal would have a minimum deviation angle of 8 degrees. A randomly oriented collection of such crystals would produce a halo with its inner edge 8 degrees from the sun. Similarly, rays taking path B through the crystal would give rise to a 17-degree halo. To explain halos of other radii we need only pick the correct angle for the pyramidal end face. The problem with an explanation that can automatically explain a halo of any size is that, in absence of some other information on the ice-crystal angles, the explanation is convincing for none of them. There is limited observational evidence on pyramidal faces in ice crystals: A few such crystals have been observed, but in general, the angles have not been determined accurately. Theory gives a number of possible angles but no clear prediction of what to expect. The angles that seem best to fit the evidence are 25 and 28 degrees.

Using the collected information from the 1974 halo display, Goldie, Meaden, and White concluded that the pyramidal angle of

Table 2-1. *Predicted halo radii resulting from light passing through pyramidal ice crystals of the form shown in Figure 2-38*

Prediction number	Entrance and exit faces	Angle between faces (degrees)	Minimum-deviation halo readings (degrees)
1	A_1B_4	28	9
2	A_1C_3	52	18
3	A_1A_4	56	20
4	B_1B_3	60	22
5	D C	62	23
6	A_1B_3	64	24
7	A_1A_3	80	35
8	D B	90	46

The pyramidal faces are designated by the letters A and C, the side faces by B, and flat end faces by D. The specific faces are designated by numbering consecutively around the crystal and giving these numbers as subscripts (see Figure 2-38). The calculations are done for a pyramidal angle (α in Figure 2-37) of 28 degrees and an index of refraction of 1.312.
Source: E. C. Goldie, G. F. Meaden, and R. White, "The Concentric Halo Display of 14 April 1974," *Weather 31*, 304 (1976).

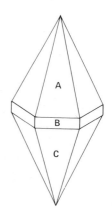

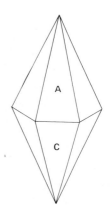

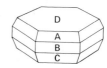

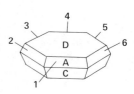

Figure 2-38. Possible forms of crystals with pyramidal faces. The numbering and lettering of the faces are used in Table 2-1.

28 degrees fit their observation better. If crystals of several pyramidal angles are needed to explain the six halos reported, the explanation is not very convincing; if one crystal shape will explain all six, it becomes more convincing. Goldie and his colleagues assumed that if pyramidal faces develop on a hexagonal ice crystal, the crystal forms of Figure 2-38 could result. They considered the possible combinations of entrance and exit faces for light rays passing through these crystals. For each combination they calculated the angle between the faces and the minimum-deviation radius of the halo that would result from rays passing through a random distribution of such pyramidal ice crystals. Table 2-1 shows the result. These investigators associated the six observed halos with the predictions numbered 1, 2, 3, 4, 5, and 7 in Table 2-1.

In my experience, the 22-degree halo on photographs frequently appears to be about 1.5 to 2.5 degrees in width, which makes it very difficult to separate halos whose radii differ by only a degree or two. From other data given by Goldie, Meaden, and White it appears that the pyramidal angle of 59 degrees would explain the six halos with nearly as good a match, and an angle of 62 degrees would again do nearly as well. Although it seems likely that the pyramidal-crystal explanation may be the right one for several of these rare halos (and Goldie and his colleagues may have deduced the correct angle), it is not clear that the last word has been said on the matter.

RARE HALO NOMENCLATURE

Besson[20] has suggested a series of names for these rare halos, each identifying a halo by the name of the first observer who reported it. The precedent for such a scheme is accepted in such names as the Parry arcs or Lowitz arcs, but I question its utility in this circumstance. As is illustrated in Table 2-1, several halos may exist with radii differing by a few degrees or less, and most observations made without photographs are not accurate enough to distinguish one from another. I also have an objection to any scheme that requires the memorization of a list of names bearing no descriptive connection to the effects they are to identify. I suggest, instead, the more prosaic solution of describing the halo by its radius: the 22-degree halo, the 18-degree halo, and so on.

TWO HALO PUZZLES

I give you Plates 2-24 and 2-25 as puzzles. A number of people who have been in Antarctica in the southern summertime have asked me about the "sun dog" that appears below the sun. Plate 2-24 shows the bright spot just at the horizon below the sun. How would you explain it?

Plate 2-25 shows a landscape on a chilly autumn morning, before the first snows of winter. The red, nail-shaped spot at the bottom of the picture is an artifact of the photograph, a ghost image of the sun resulting from reflections within the camera lens. But the sparkles of light in the foreground are real — and caused by what? Try first your answers and then mine, from the Appendix.

3

Ice-crystal reflection effects: pillars, circles, and crosses

SUN PILLARS

Puzzlement over how sun pillars are formed led me to seek the answer in computer simulations. I was explaining sun pillars to an especially interested student and describing the vertical column of light that can sometimes be seen extending above and/or below the sun when it is low in the sky. As my hearer was following closely, I continued to explain in careful detail why a sun pillar has the shape that is observed. The more careful I became, the more slowly the explanation proceeded, until it ground to a halt as I realized that I did not understand the matter myself. I had been aware of sun pillars for several years, but until that time, I had not realized my own ignorance. The common explanation has been that flat-plate crystals fall in still air so that their flat bases are oriented nearly horizontally. Reflection of sunlight off the slightly tipped surfaces is said to produce the long column of light shown in Plate 3-1. Minnaent[1], in his delightful book *The Nature of Light and Colour in the Open Air,* discusses this explanation and its shortcomings and ends with the statement: "The pillars of light seemed such a simple phenomenon. Who would have thought that their explanation would incur so many difficulties?"

The student and I realized that we could test the prediction of the simple model proposed using the computer to calculate the direction a ray from the sun takes after it is reflected from a nearly horizontal crystal face. As described in Chapter 2, the process involves assigning the crystal a large number of different orientations and representing the result of each reflection as a spot. The spot indicates the direction in the sky from which light would come to the observer's eye from a crystal with that particular orientation, and the resulting spot picture should show the shape of the light pattern that would be seen in the sky. We need only choose the appropriate distribution of orientations to be able to proceed.

Figure 3-1 illustrates a convenient way to talk about these orien-

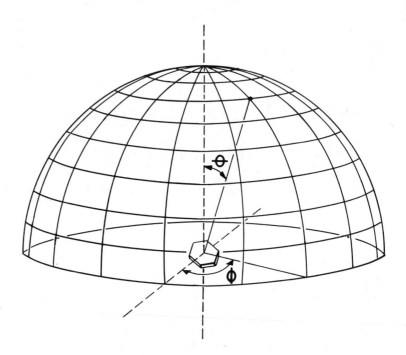

Figure 3-1. The distribution sphere used to describe the orientations of the reflecting faces of plate crystals.

tations. The reflecting surface is placed at the center of a sphere inscribed with latitude and longitude lines. Consider a line perpendicular to the reflecting face. The point at which that line intersects the sphere specifies the orientation of the crystal face. For example, a point at the north pole represents a horizontal face; a point on the equator represents a vertical face. In this diagram θ represents the tilt angle of the crystal, and ϕ represents the azimuthal angle (the angle measured in the horizontal plane, as shown in Figure 3-1).

SUN PILLARS FROM FLAT-PLATE CRYSTALS

For our first attempt at computer simulation of a sun pillar,[2] we chose a collection of surfaces that had a uniform distribution of tilt angles up to a maximum angle, θ_{max}. By a *uniform distribution* we mean one that can be represented by a uniform density of points on the sphere in Figure 3-1. Obviously this distribution is a simplified one: We could have chosen to taper off the edges rather than to select an abrupt θ_{max} cutoff. However, the essential predictions of this model should emerge either way; so we chose to look at the simplest case.

Figure 3-2 shows the pattern resulting from a program specifying the sun at 4 degrees above the horizon and a θ_{max} of 6 degrees. The sun is drawn to scale, and the horizon is represented by a heavy line. Remember that the portion of the pattern below the horizon can be seen by an observer in an airplane above a cloud of ice crystals.

The program that produced the form shown in Figure 3-2 ignores several factors. One of these is the finite size of the sun; the pattern is that predicted for a point sun. To take account of the sun's 0.5 degree angular diameter we could take the pattern and extend it by 0.25 degree in all directions. The program also ignores the effect of the projected area of the crystal face on the intensity of the reflection; for example, at grazing incidence (where the ray is nearly parallel to the crystal face), the reflected beam of light has a small cross-sectional area and, hence, a low intensity. We partially compensated for this effect by neglecting the fact that the reflectivity of the ice surface increases for rays with more nearly grazing incidence. There is another effect that would complicate the intensity calculations: Light entering a side face of the crystal can experience total internal reflection on an end face and emerge from the opposite side face in the same direction as the ray reflected from the top face (Figure 3-3).

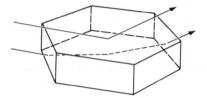

Figure 3-3. Two ray paths that can contribute intensity to the sun pillar.

The intensity of such rays may be high because of the total internal reflection. Calculation of the intensity of these rays would involve the aspect ratio (the ratio of thickness to diameter) of the crystal. Despite these omissions, the essential features of the model prediction can be seen.

The regularity of the pattern shown in Figure 3-2, displayed as lines and arcs of dots, is clearly an artifact of the method of calculation. In our first computer program we changed the values of θ and ϕ from one calculation to the next by regular increments in order to keep a constant density of points on the distribution sphere. It is this regular set of angle increments that produces the structure shown in Figure 3-2. Later we randomized the choice of angles to remove the spurious patterns.

Light reflected from slightly rough water is another situation in nature where you see a light pattern resulting from reflection by an array of nearly horizontal surfaces. The simulation shown in Figure 3-2 represents the shape of a glitter path seen on the water as the sun is about to set. Compare the computer simulation with Plate 3-2. You can see how the combination of photograph and computer simulation could be used to determine the slopes of the wave surfaces.

Figure 3-4 shows the predicted shapes of sun pillars produced by flat-plate crystals for sun elevations ranging from 6 degrees above

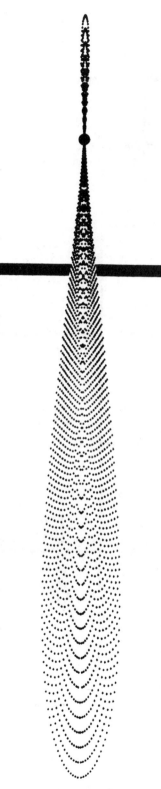

Figure 3-2. Simulated sun pillar from flat-plate crystals.

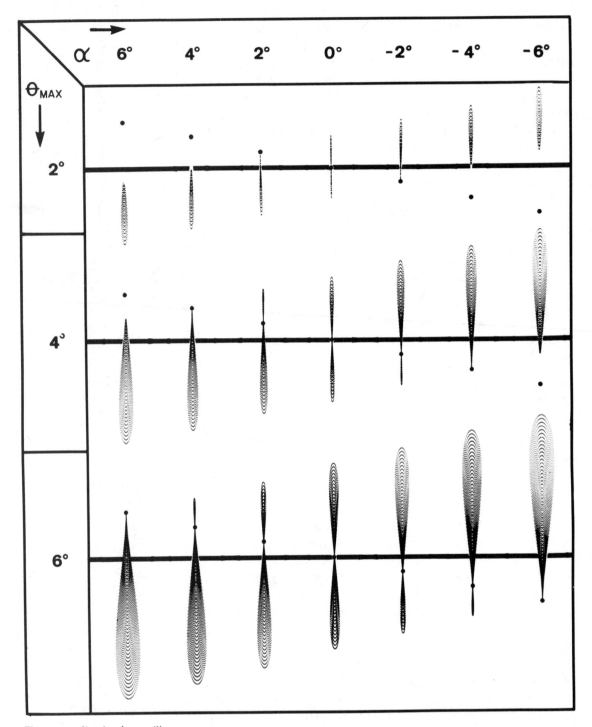

Figure 3-4. Simulated sun pillars from flat-plate crystals. A horizontal sequence shows how the pillar changes as the sun sets. Each of the three sequences represents a different degree of ice-crystal orientation. the horizon ($\alpha = 6$ degrees) to 6 degrees below the horizon ($\alpha = -6$ degrees). The resulting sequence is given for three different values of θ_{max}. In the figure, the horizon is represented by a heavy horizontal line and the sun is shown as a disk drawn to scale. From these

sequences we can make several predictions about the appearance of light pillars formed by this mechanism. The larger maximum-tilt angles (poorer orientations) produce pillars of greater vertical extent and of greater width. Note that the greater the elevation of the sun, the greater the tilt angles required to produce a pillar near the sun. Apparently, flat-plate crystals are most likely to produce pillars when the sun is low in the sky or even after it has sunk below the horizon.

SUN PILLARS FROM PENCIL CRYSTALS

Figure 3-4 deals with sun pillars that *result from reflection from the surfaces of flat-plate crystals.* Sometimes a short pillar is seen passing through the sun or moon, extending above and below it for sun elevations greater than 10 degrees. That shape is not predicted by the simulations shown in Figure 3-4. But such a pattern can be produced by reflections from the same kinds of crystals that produce the circumscribed halo (i.e., pencil crystals falling with their long axes horizontal). One side face of a pencil crystal can assume every possible orientation and so can reflect light in every direction. This is in contrast to the slightly tilted flat-plate crystal, which can reflect light only in certain limited angular directions. The lack of such limited reflection directions is probably the reason no one had previously suggested that pencil crystals might produce sun pillars. But computer simulations of reflection from the sides of pencil crystals produced the results shown in Figure 3-5. Although reflected light

Figure 3-5. Simulated sun pillars from pencil crystals. The sequence shows how the pillar changes as the sun sets. The sun position is represented by a white disk.

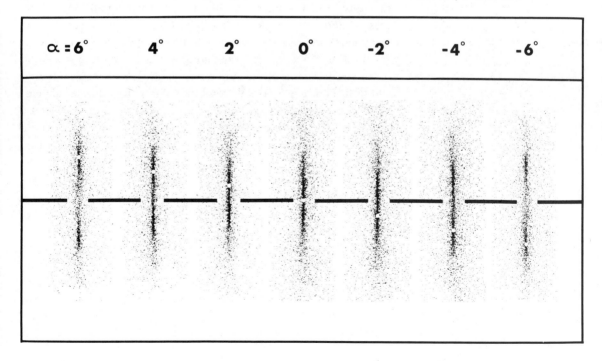

$\alpha = 6°$ $4°$ $2°$ $0°$ $-2°$ $-4°$ $-6°$

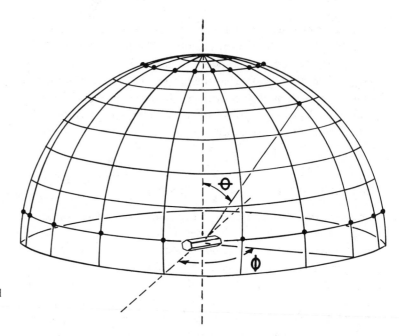

Figure 3-6. Distribution sphere used to describe the orientation of pencil crystals (with long axes horizontal).

can appear from all directions, there is a significant concentration in the direction of the sun, and a pillar of light extending above and below the sun appears in each simulation.

Hindsight is often much clearer than foresight. Knowing the answer does indeed make it easier to predict what the answer should be. So it is with pencil-crystal sun pillars. Think about the distribution of orientations of the crystal faces using the distribution sphere in Figure 3-6. Consider the case of exactly horizontal long crystal axes, assuming that the crystals will have all rotations about the horizontal axes and that the axes may point in any direction in the horizontal plane. Note that this does not result in a uniform distribution of tilt angles for one of the long crystal faces. For example, if both θ and ϕ in Figure 3-6 take all combinations of the values o degrees, 10 degrees, 20 degrees, and so on, the orientations of the crystals are represented by points at each intersection of a "latitude" line and a "longitude" line on the distribution sphere. Such a distribution results in a concentration of points around the north pole of the sphere, and it is this concentration of horizontal surfaces that gives rise to the pencil-crystal sun pillar!

One reason for questioning the flat-plate origin of sun pillars was the difficulty of explaining the occasional pillar that appears to pass through the sun when the sun is 10 or 15 degrees above the horizon. Figure 3-7 shows the predicted pattern of reflection from the surfaces of pencil crystals for sun elevations of o, 10, and 20 degrees. At 20 degrees the pillar is still distinguishable, although it has a small vertical extent. Note that, as depicted in Figure 3-5, the concentration of light is centered on the sun at each elevation; this agrees with

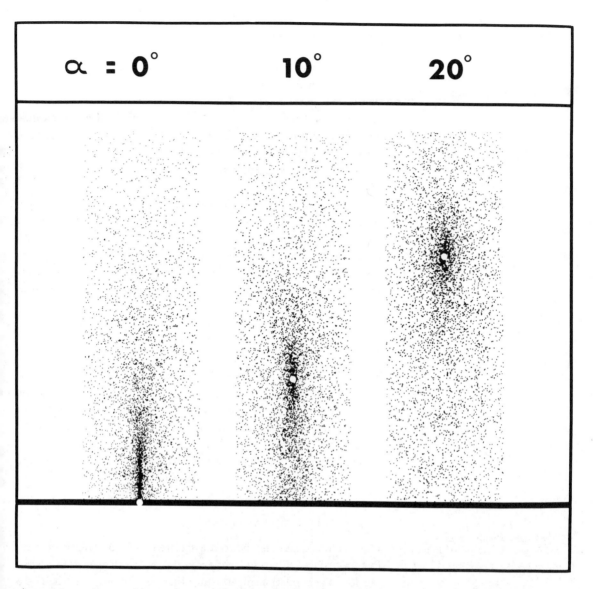

$\alpha = 0°$	10°	20°

observations that could not be explained by the flat-plate-crystal hypothesis.

Sometimes the source of a sun pillar can be identified by comparing the pillar's shape with the predictions of the two models. This is not always possible, however. The shapes do not differ greatly when the sun is near the horizon, and nonuniform cloud layers can emphasize a part of the pattern while diminishing other parts. Sometimes (Plates 3-3 and 3-4) the refraction effects that occur simultaneously with the sun pillar serve as diagnostic features to identify the ice crystals producing the pillar.

Plate 3-3 shows a fairly high sun pillar and an elongated sun dog located 22 degrees to the left of the sun. Both these features lasted for more than half an hour as the clouds drifted across the sky; so it is

Figure 3-7. Simulated sun pillar from pencil crystals for sun elevations of 0, 10, and 20 degrees.

7 1

likely that both resulted from the same crystals in the rather uniform cloud layer. For the computer identification we selected a distribution of flat-plate crystals with tilt angles of up to 17 degrees, necessary to produce a pillar of the height observed with the sun 10 degrees above the horizon. Then, *with this same distribution of ice crystals,* we simulated the 22-degree refraction effect to see if it would agree with the elongated sun dog in the photo. The simulation is shown next to the photograph. Agreement is good enough to suggest that this sun pillar was indeed produced by flat-plate crystals.

Plate 3-4 shows a sun pillar surmounted by an arc, both thought to have a pencil-crystal origin. The simulation shows both reflection and refraction from the same distribution of pencil crystals with long axes horizontal and random rotational orientation about their axes. The simulation agrees in surprising detail with the photograph, leaving little doubt that this sun pillar was produced by pencil crystals.

MOON PILLARS

The light pillars discussed above can be associated with either the sun or the moon. I recall one crisp evening many years ago when, on the last run down a ski slope, I watched the rising moon's shimmering column of light, formed by myriad crystals sparking in the moonlit air. It was the first pillar – sun or moon – I had ever seen. I wonder if my vivid memory of the beauty of that scene sparked my later interest in understanding these effects.

ARTIFICIAL LIGHT PILLARS

Figure 3-8. The difference in the height of light pillars formed by (A) sunlight and (B) a local light source.

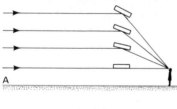

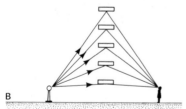

Light columns can also be seen associated with local light sources. Plate 3-5 shows pillars of light extending above the lamps in a parking lot. Their relationship to sun pillars is obvious, but there is a significant difference: Sun pillars are seldom seen extending as much as 20 degrees above the sun, but light pillars from nearby sources can be much higher. The pillars displayed in Figure 3-4 have heights that increase with the tilt of the flat-plate crystals. The total angular extent of the light pattern, from top to bottom, is four times the maximum-tilt angle of the crystal distribution; for example, if the maximum-tilt angle is 1 degree, the total vertical extent of the resulting pillar is 4 degrees. As the crystal tilt becomes greater, the pillar appears broader and more diffuse, and the fainter parts of the predicted patterns are likely to be invisible. Figure 3-8A illustrates how the crystal tilt needs to increase to reflect the distant sunlight to the eye from crystals higher in the sky (i.e., from higher parts of a sun pillar). Figure 3-8B shows how crystals with no tilt

can reflect light from a nearby source to the eye from any height in the sky. Even with untilted crystals, the pillar associated with a local light source can, in principle, have an infinite height and can be very narrow. Figure 3-9 shows narrow artificial light pillars extending above streetlamps on a cold night in Wyoming.[3]

Figure 3-9. Artificial light pillars above streetlights near Laramie, Wyoming. (Photographed by Kenneth Sassen)

One evening in a rural area where there were not many lights, I saw a particularly strong moon pillar. The next day I learned that the nearby city had been dazzled by an unusual auroral display featuring columns of light converging to the zenith. I am quite sure that what the townsfolk really saw was a forest of high vertical light pillars, each originating from a city light; the apparent convergence to the zenith was the perspective effect of parallel lines.

THE SUBSUN

If the sun is high enough in the sky and the flat-plate crystals are well enough oriented, the pillar takes the form of an elongated spot lying entirely below the horizon. For fairly high sun elevations, the spot bears so little resemblance to a sun pillar that it has its own name, the subsun. Compare the upper left simulation in Figure 3-4 with the photograph in Plate 3-6. The subsun shown in the photo-

73

graph is produced by ice crystals so near that you can see that the effect is formed by individual reflections, giving individual spots of light. (I am tempted to say that this photo is really a good simulation of our computer spot diagram.)

As the crystal faces become more nearly horizontal, the sun pillar below the horizon shrinks in vertical dimension. In the limit of perfect orientation, the spot becomes a reflection of the sun in the sea of horizontal mirrors. It is located as far below the horizon as the sun is above. Plate 3-7 shows a brilliant subsun below the wing of the airplane from which it was photographed. You can see such a display fairly commonly as you fly over ice-crystal clouds, and it is sometimes of dazzling brilliance. The subsun spot moves along with the airplane through its climbs, dives, or turns, but disappears when the ice clouds vanish. I suspect subsuns account for at least some of the things reported as UFOs.

If you watch the subsun formed in a thin cloud layer, through which you can see the ground, you will see that as it passes over a lake, pond, or river, the subsun flashes with the brilliance of the sun reflected from the horizontal water surface – just what you would expect from the foregoing explanation. The vertical dimension of the spot supplies a measure of the orientation of the crystals. I conclude that most of the ice crystals contributing to the subsun shown in Plate 3-7 have tilts smaller than 1 degree. This seems a surprising degree of orientation, but, judging from the appearance of subsuns, it is not unusual.

THE PARHELIC CIRCLE

As described previously, there are two ways that nearly horizontal surfaces can occur when small ice crystals fall through the atmosphere: (1) by the selective orientation of flat hexagonal plates, with their bases nearly horizontal, and (2) by the statistical concentration of nearly horizontal side faces of pencil crystals with their long axes horizontal. In both these cases, the ice crystals also have faces selec-

Figure 3-10. Reflected rays from vertical surfaces give rise to the parhelic circle.

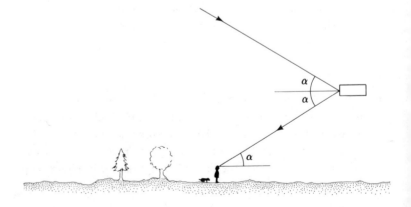

tively oriented vertically; these are the side faces of the plate crystals and the end faces of pencil crystals. What would you expect to result from sunlight reflected from these vertical faces? A vertical mirror that intercepts a ray of light coming from the sun 30 degrees above the horizon reflects the light in a direction 30 degrees below the horizon. An observer would look up to see the ray coming from a direction 30 degrees above the horizon (Figure 3-10). As the vertical mirror addresses different directions around the horizon, it reflects the light in different azimuthal directions, but always directs the light downward at an angle of 30 degrees. The net result of a collection of vertical mirrors with all other possible orientations is a circle of white light extending around the sky at constant elevation above the horizon, passing through the sun. Such an effect is observed and is called the parhelic circle.

Plate 3-8 shows a display with the sun behind the globe of a streetlamp. You see what appear to be parts of the 22-degree halo and of the circumscribed halo. The nearly horizontal curved line running off to the left of the sun is the parhelic circle. If you expected it to appear as a straight line, remember the discussion in Chapter 2 concerning sun dogs and circumzenithal arcs. To verify the identification of the three arcs shown in the photograph, we have done a computer simulation of the two halos for a sun elevation of 50 degrees, giving the same perspective as that of the photograph (Figure 3-11). On this simulation is included a constant-elevation line marking the position of the parhelic circle. I would say that the match of three arcs in the simulation and the photograph confirms the diagnosis.

Also included in the simulation is a weak sun dog, which matches a faint feature just outside the circumscribed halo in the photograph. Because the white tail of the sun dog lies superimposed on

Figure 3-11. Simulation of 22-degree halo, circumscribed halo, and the parhelic circle of Plate 3-8.

the parhelic circle, the two elements might be confused. A clear identification can be made by determining whether the line extends inside the halo — where it can only be part of the circle. Sometimes the circle can be seen extending completely around the sky. Extended parhelic circles can be seen in some of the photographs taken with a fisheye lens that are presented in Chapter 4.

COMBINATIONS OF REFLECTION AND REFRACTION

In nature, when sunlight encounters a group of ice crystals falling in the sky, every reflection and refraction that is possible occurs. To sort out and understand the amazing array of effects that can (and do) result when sunlight communicates with a simple hexagonal prism of ice, I have attempted to look at the effects in a systematic fashion — to understand them one by one, hoping in the end to view nature's wholeness with an appreciation that comes from intimate acquaintance with each of its parts. In Chapter 2 I discussed the passage of light through the ice crystal and how it is refracted upon entering and upon leaving. So far in this chapter I have discussed a single reflection from a crystal face. The next step is to look at combinations of reflection and refraction to see what they contribute to the sky picture.

TWENTY-TWO-DEGREE SUBPARHELIA (SUBSUN DOGS)

You can occasionally see from an airplane the subsun flanked on either side by sun dogs, which are appropriately called 22-degree subparhelia or subsun dogs. It is easy to imagine that the bright subsun is acting as a secondary source to produce its own parhelia. I think this is unlikely to be the correct explanation; in fact, I am suspicious of most explanations that involve rays of light interacting successively with two different crystals, because the intensity of such secondary phenomena would be very low. Although a subsun may appear bright, it is still several orders of magnitude fainter than the sun. If the reflected subsun light were the source for the subsun dogs they should be fainter than ordinary sun dogs by the same large factor. According to my observations, they can be of comparable intensity.

Plate 3-9 shows a subsun and its 22-degree companion taken, not from an airplane, but while looking down into ice crystals in a river valley. Both of the features are elongated from the tilts of the flat-plate ice crystals producing them.

If the sun is near the horizon, some of its rays may enter one side of a thin plate crystal, pass through the crystal, and exit through an alternate side face, as described earlier. But when the sun is at a great enough elevation above the horizon, all the rays that enter a

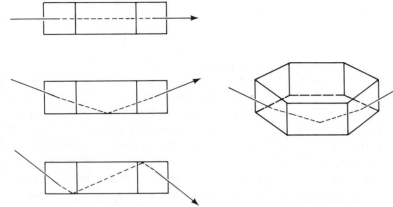

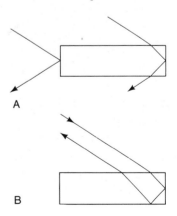

Figure 3-12. The paths of rays that contribute intensity to the sun dogs and subsun dogs.

side face of a thin crystal will strike the flat end (bottom) face. After reflection (Figure 3-12), the rays may exit through the alternate face and produce a sun dog, but one seen below the horizon, reflected in the same way as is the subsun. The ray will strike the bottom face near grazing incidence and for a wide range of sun elevations will be totally internally reflected. In fact, in a well-developed plate crystal, more than one internal reflection could occur with no light loss. Any odd number of reflections would contribute to the subsun dogs; any even number of reflections would add intensity to the normal parhelia. I think it likely that the normal parhelia, occurring for sun elevations above 20 or 30 degrees, derive most of their intensity from such even-numbered multiple internal reflections.

The once-reflected ray of Figure 3-12 makes an angle with the horizontal, exactly as if it were externally reflected from the upper face. As a consequence we expect the vertical extent of the subsun dogs to be nearly the same as that of the subsun – as in Plate 3-9.

SUBPARHELIC CIRCLE

The parhelic circle, which results from light reflected by vertical surfaces, could equally well result from either external or internal reflections (Figure 3-13A). In fact, for sun elevations greater than 32 degrees, the internal ray is totally reflected, and these rays can be expected to contribute significantly to the intensity of the parhelic circle. Following an internal reflection from the side of a plate crystal, the ray can be reflected from the bottom face (Figure 3-13B) and emerge from the top of the crystal, making the same angle with the top face as the incident ray. Such twice-reflected rays could produce a parhelic circle, reflected so that it appeared below the horizon. The result would be a subparhelic circle visible from an airplane, extending around the sky at a constant angle below the horizon. The argument would be the same if Figure 3-13B represented a side view of a pencil crystal with its long axis horizontal and one pair of the side faces also horizontal, the orientation that is responsible for the

Figure 3-13. The paths of rays that contribute intensity to the parhelic circle and the subparhelic circle.

A

B

Parry arcs. The subparhelic circle shows in Plate 3-13, along with another set of arcs that I will discuss shortly.

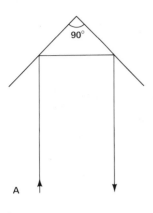

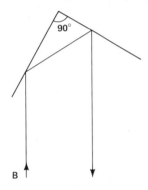

Figure 3-14. Reflection from two mirrors positioned with a 90-degree angle between them.

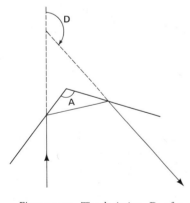

Figure 3-15. The deviation, D, of a ray reflected from two mirrors positioned with an angle, A, between them.

EFFECTS SEEN FAR FROM THE SUN

The side of the sky away from the sun has not been well explored for visual effects. Part of the reason may be our inclination to concentrate on the spectacular things near the sun and not to look for fainter, poorly understood effects in the opposite direction. There are at present numerous one-of-a-kind sketches of such phenomena but remarkably few photographs. It seems that most of these effects require two or more internal reflections, and their analysis becomes quite complicated: When we consider all possible combinations (for a hexagonal ice prism) of entrance face, two or three internal reflection faces, and an exit face, we have a very long list to consider. Before attempting to deal with the observed effects, let us consider a principle that will be quite helpful in reducing the length of this list.

REFLECTION FROM TWO MIRRORS

Figure 3-14 shows a ray of light approaching two mirrors, positioned with a 90-degree angle between them; the mirrors are perpendicular to the plane of the page, and the ray lies in the plane of the page (the normal plane). If the ray strikes the first mirror at an incident angle of 45 degrees, it strikes the second mirror also at 45 degrees and leaves the pair of mirrors parallel to its approach direction (Figure 3-14A). If the pair of mirrors is rotated so that the angle of incidence on the first is 60 degrees (Figure 3-14B), the ray strikes the second mirror at 30 degrees and again emerges parallel to the incident direction. In fact, it is not difficult to show that, however the pair of mirrors is turned, as long as the angle between them is 90 degrees, a ray in the normal plane will emerge after two reflections parallel to its incident path. Because of this property, such a pair of mirrors acts as a very interesting looking glass. If you look into the pair you see a reflection of yourself that appears, on first glance, to be what you would see in a usual plane mirror. But the pair has the unusual property of enabling you to stand anywhere within the 90-degree angle and see your reflection. Each of several people standing side by side in front of a pair of relatively small mirrors will see his or her own reflection. With that useful feature, why isn't such a system used in public places to save mirrors? If you attempt to comb your hair you will know the answer. With a plane mirror, when you lift your right hand, your image behind the mirror lifts its left. Not so with this device; here your image lifts its right hand, on the other side of the mirror. It makes it difficult to comb the hair.

We can generalize this result for reflection from two mirrors in terms of the angle between the two mirrors (A) and the total change in the direction of the beam resulting from the two reflections (the deviation angle, D). These angles are defined by Figure 3-15. For the case we have been discussing, A = 90 degrees, the deviation angle D = 180 degrees. For each angle A, there is a value of D for rays in the normal plane that does not depend on the rotation of the pair of mirrors. The relationship between the two angles is

$$D = 2(180° - A)$$

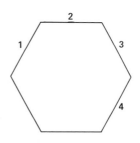

For internal reflection in a right hexagonal ice prism, there are only a few possible values of A. Between a side face and an end face, the angle is 90 degrees, the case I discussed first. In Figure 3-16, all the other possible combinations of two reflections are shown. Between adjacent faces (1 and 2), A = 120 degrees and the deviation angle is 120 degrees, as shown. Between alternate faces (1 and 3), A = 60 degrees and D = 240 degrees. For opposite faces (1 and 4), A = 180 degrees, with a resulting D of 360 degrees. This means that the ray, after its encounter with the pair of mirrors, continues in the same direction as before; A deviation of 360 degrees is the same as a deviation of 0 degrees, which is no deviation.

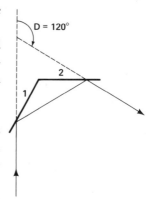

There is an additional bit of insight to be gained from Figure 3-16 that will prove useful later. To see light that has been deviated by 180 degrees, one must look directly away from the sun (toward the antisolar point). Reflections off adjacent faces and off alternate faces give deviations of 120 and 240 degrees, respectively, but Figure 3-16 demonstrates that both these deviations result in light making an angle of 60 degrees from the straight-back direction. To see light reflected from either of these pairs of faces, you have to look in a direction 60 degrees from the antisolar point. In many cases, then, the result of both sets of reflections is the same. This idea significantly reduces the number of combinations we need to investigate.

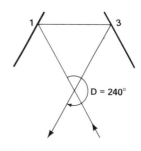

ONE HUNDRED TWENTY-DEGREE PARHELIA

Of the relatively rare effects seen in the sky away from the sun, the least rare are probably the 120-degree parhelia, bright spots that appear on the parhelic circle 120 degrees away from the sun. It is convenient to introduce a term to describe a point on the parhelic circle opposite the sun: the *anthelic point,* not to be confused with the antisolar point, which, for the sun above the horizon, will lie opposite the sun and below the horizon. The anthelic point lies 180 degrees in azimuth away from the sun (180 degrees measured in a horizontal plane), but has the same elevation above the horizon as the sun. (If you like to play with words, you might like to consider that another correct name for the antisolar point would be the *subanthelic point.*) An equivalent way to describe the position of the 120-degree parhe-

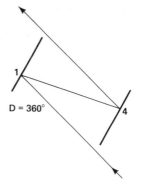

Figure 3-16. Three possible two-mirror reflection combinations in a hexagonal prism.

79

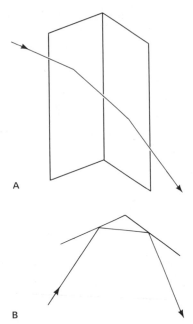

A

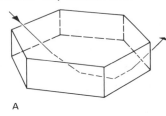

B

Figure 3-17. Two-mirror reflection for a skew ray. A: Perspective view. B: Top view.

Figure 3-18. Two light ray paths that can contribute to the 120-degree parhelia. A: Ray enters top face and exits from bottom face after two internal reflections. B: Here there is transmission and reflection only from side faces.

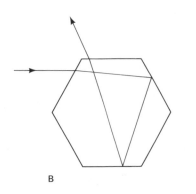

A

B

lia is to say they lie on the parhelic circle 60 degrees in azimuth from the anthelic point. Their proximity to the anthelic point is sometimes emphasized by referring to them as paranthelia, but I prefer to call them 120-degree parhelia. Following the earlier discussion, we can see that light at these positions can result from reflection from two vertical mirrors with an angle between them of either 120 degrees or 60 degrees. For that explanation to be convincing, I need to make an additional argument.

The two-mirror discussion was previously restricted to rays in a normal plane; that is, for two mirrors that are both vertical, to rays in the horizontal plane. But now I must consider rays that come sloping down from the sun when it is above the horizon. Figure 3-17A shows a perspective drawing of the situation for reflecting planes 120 degrees apart, and 3-17B shows it in top view, with the paths of the rays projected on a horizontal plane. The angles of incidence and reflection, as measured on this projection in Figure 3-17B, are not the actual angles of incidence and reflection measured on the mirrors in 3-17A, but all the laws of reflection are obeyed on the projection; that is, the incident and reflected angles as measured on the projection are equal. Therefore, for the 120-degree mirrors, the deviation *in the horizontal plane* is 120 degrees. The net result of this double reflection is that the ray is deviated 120 degrees in azimuth but continues sloping downward at the same angle. To see such a ray, you look upward at the elevation of the sun, but 120 degrees away from the sun in azimuth — and that is just where you see the 120-degree parhelia.

But yet another detail must be treated. Internal reflection from two faces, either adjacent or alternate, will give the correct deviation to produce the 120-degree parhelia. But how can the light get into the crystal and then out of it without introducing more deviation? From most entrance and exit faces, there is indeed additional deviation, by an amount equal to that introduced by a 60-degree prism. There are two possibilities, however, for which this is not the case. Figure 3-18A shows a ray entering the top face, reflecting off two internal faces, and exiting from the bottom face. Figure 3-18B shows a case where the additional deviation introduced by refraction at the entrance face is exactly compensated for by the refraction at the exit face. Either mechanism (or possibly both) could produce the 120-degree parhelia, and for either, an additional reflection off the bottom face could produce a 120-degree subparhelion, though I know ·of none that has been reported.

EFFECTS AT THE ANTHELIC POINT

The center for reported activity on the side of the sky away from the sun is the anthelic point. Diagonal arcs passing through the point,

Figure 3-19. Anthelic arcs for sun elevation of 50 degrees. (Photographed by A. James Mallmann)

marking the spot with a dramatic X, have been reported. And, although a few photographs have been obtained (Figure 3-19 and Plate 3-10),[4] until recently that X might have represented an unknown factor: the mechanism by which these anthelic arcs are produced. Other anthelic observations include a concentration of light at this point (the anthelion) and a vertical column of light that I will call the anthelic pillar. Plate 3-11 shows an anthelic pillar, and Plate 3-12 shows a cusp feature with its tip at the anthelic point.

Several of the suggestions made to explain these effects suffer from lack of quantitative predictions. My colleagues and I wanted to use our computer-simulation technique to explore the consequences of the variety of suggestions for anthelic phenomena, but we did run into some difficulty. It was a straightforward task to modify our computer program so as to trace the direction of a ray as it reflects internally from any of the crystal faces and hence to determine its direction after any number of internal reflections. In simulating the effects that result when rays are refracted through the ice crystals, we included intensity calculations that account for reflection losses of light intensity and also for the restriction of the cross section of the transmitted beam by the finite size of each crystal face. We had to work rather hard to get the correct general intensity factor for the

81

refraction problem, one that would hold for any arbitrary orientation of the crystal. The geometric factors – the limitation imposed by the boundaries of the entrance and exit faces – were the difficult part. To these we wanted to add the additional complications of one or two or three additional faces, providing internal reflections, between the entrance and the exit faces. Without some clever insight, which we did not have, the problem seemed too complicated; we decided to do the calculations without any intensity factors.

Accordingly, we tried to draw conclusions from the geometry of the effects (e.g., the angle between the anthelic arcs) rather than from the intensity distributions. For several of the suggestions offered to explain anthelic effects, no quantitative predictions had previously been made. The danger in our process is this: We may predict things that do not occur. Our program calculates the direction of a ray after it reflects from a plane with the orientation of one of the crystal faces. It may be that some rays do not get to that face because they hit other faces first; but our calculation includes their contributions nevertheless. As a result, we may predict that arcs extend farther than they actually do, or plot things that are of too low intensity to be seen, or even plot some patterns that cannot physically occur. However, we should not miss any effect that does occur. The significance of the limitation is that our results,[5] though they may be quite useful for explaining observed effects, are less useful for predicting new effects that have not been observed.

ANTHELIC ARCS

Hastings,[6] in 1920, and Wegener,[7] in 1925, suggested quite similar mechanisms to explain the anthelic arcs. Both considered rays that pass through a pencil crystal with its long axis horizontal, entering a side face and exiting from an alternate side after an intermediate internal reflection from the end of the crystal (Figure 3-20). Their suggestions differ only in the assumed orientation of the crystal; Wegener assumed that the crystal could have any degree of rotation around its horizontal axis, whereas Hastings assumed that a pair of the side faces should be horizontal. I have discussed both of these distributions before; the first is the circumscribed-halo distribution and the second, with the additional degree of orientation, is the Parry-arc distribution. The consequences of these suggestions are shown in the first two columns of Figure 3-21. As might be expected, the results are similar, the more restricted orientations (Hastings's) giving the narrower set of arcs.[8] Both sets look like plausible candidates to explain the photographs of Figure 3-19 and Plate 3-10.

One obvious measure of the fit of the predictions is the angle be-

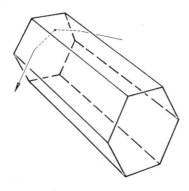

Figure 3-20. Path of a ray that contributes to the anthelic arcs of Wegener or of Hastings.

	WEGENER	HASTINGS	TRICKER

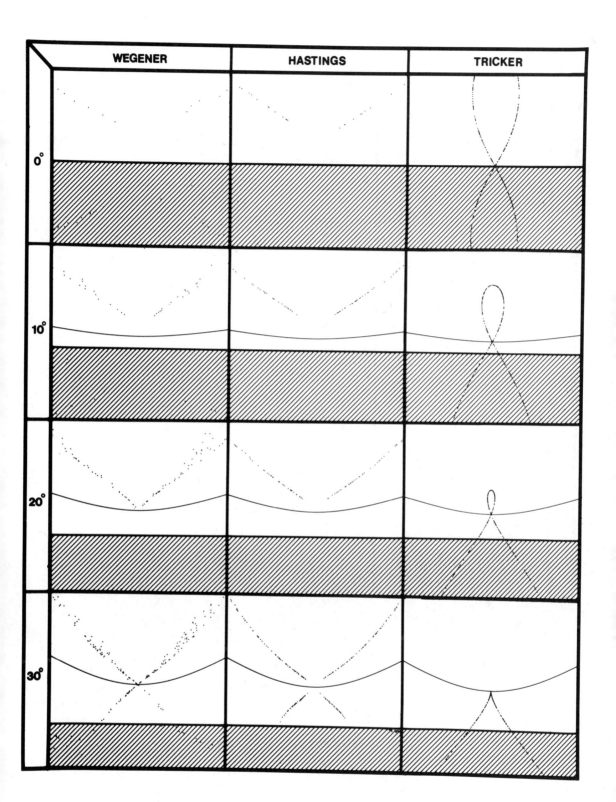

0°

10°

20°

30°

8 3

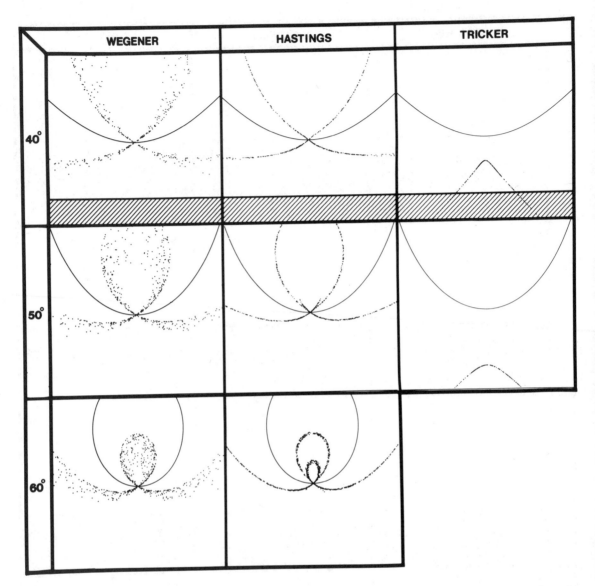

	WEGENER	HASTINGS	TRICKER
40°			
50°			
60°			

Figure 3-21. Simulations of three different kinds of anthelic arcs.

tween the arcs. I define the anthelic-arc angle (Figure 3-22) as the vertical angle between the intersecting arcs. The anthelic arc display of Plate 3-10 occurred for a sun elevation of 38 degrees and that of Figure 3-19 for 50 degrees. We measured the arc angles for those photographs, to compare them with the values measured from the simulations. We repeated the measurements several times, using different methods, to get a sense of the uncertainty involved. For the 38-degree elevation (Plate 3-10) we obtained the measurements 114 ± 5 degrees, which agrees with the Wegener value of 111 degrees better than with the Hastings value of 106 degrees. For the 50-degree elevation (Figure 3-19) our measurement was 120 ± 5 degrees, again in better agreement with the Wegener value (116 degrees) than with the Hastings value (107 degrees). These results

84

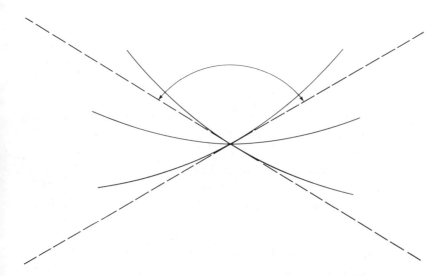

are consistent with observations made from other photographs taken at the same time as these two. In both cases, photographs of the 22-degree effects around the sun show strong circumscribed halos but no trace of Parry arcs. (Plate 3-8 came from the same display as Figure 3-19, and Figure 2-32 of the previous chapter was photographed at the same time as Plate 3-10.) It would be interesting to analyze a photo of anthelic arcs occurring while a Parry arc was visible on the other side of the sky. Because circumscribed halos are more common than Parry arcs, we should expect Wegener's anthelic arcs to be more common than Hastings's arcs. In addition, many Parry-arc observations include the superimposed circumscribed halo (or upper tangent arc); so it seems unlikely that we will see a Hastings anthelic arc that is not overlaid with a Wegener arc.

Tricker[9] has suggested another mechanism that might give rise to anthelic arcs, particularly for low sun elevations. He considers a ray that enters the end face of a pencil crystal (with horizontal axis), makes two internal reflections from the side faces, reflects off the other end face, and exits from the same end face it entered. Figure 3-23 shows such a ray. Tricker likens this multiple reflection to the reflections that take place in a kaleidoscope. Consideration of all the variations on such a scheme — different combinations of side faces with the end reflection coming before, between, or after the side reflections — at first makes the number of possibilities to investigate seem discouragingly large. In this instance, as in many others, however, additional insight into the problem drastically reduces the number of possibilities and makes it look much simpler.

When the complicated ray path of Figure 3-23A is seen as viewed from the end of the crystal, the result is Figure 3-23B; that is, 3-23B is a projection of the ray onto the normal plane, the plane of the page. In that projection the direction of the ray is not changed by

Figure 3-22. Definition of the vertical angle between intersecting anthelic arcs.

Figure 3-23. Path of a ray for Tricker's kaleidoscope anthelic arcs. A: Perspective view. B: End view.

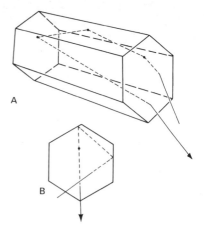

A

B

85

reflection off the far end face (marked by a dot in the diagram). So the diagram would not be changed if the end reflection occurred before the two side reflections, or between them. I have described earlier how the deviation of a ray from two mirrors depends only on the angle between the mirrors. We can conclude that the direction of the reflected ray does not depend on the order in which the several reflections occur. That idea alone significantly reduces the number of combinations we need to consider.

The ray inside the crystal, before it has undergone any reflections, makes a certain angle with the normal plane; that is, with the plane of the paper in Figure 3-23B. None of the reflections off the side faces change that angle between the ray and the plane of the paper, and in fact, reflection off the far end face merely reverses the ray, sending it back, so that it retains its initial angle with the plane. The result of all the reflections is that the ray returns, making the same angle with the normal plane as initially, but with a side component (a component in the normal plane) that is given by the construction of Figure 3-23B or, more generally, by the drawings of Figure 3-16. But remember that from a distance you cannot tell the difference between reflections from a pair of mirrors set at 60 degrees and at 120 degrees to each other; you must look 60 degrees away from the straight-back position to see the light in either case. Therefore, successive reflections from adjacent side faces or from alternate side faces give the same result, and this is another simplification. Consider now the ray that is reflected off opposite side faces in Figure 3-16. Its direction in the normal plane is unchanged by the double reflection, and so the net result, including the internal end-face reflection, is just the same as external reflection from a vertical surface; it should only add intensity to the parhelic circle.

What if we considered, instead of reflections off two side faces, reflections from four or six or any even number of side faces? Each pair of reflections from the faces of a hexagonal crystal can result in deviations of 120, 240, or 360 (0) degrees, corresponding to the three possibilities shown in Figure 3-16. Any number of successive *pairs* of reflections can produce only the sum of these angles, which all reduce to deviations of 0, 120, or 240 degrees, just as for reflections from two side faces. Accordingly, *any combination of an even number of side-face reflections, plus one end-face reflection, in any order, will produce the same set of arcs.* This real simplification certainly makes the simulation worth while.

The third column in Figure 3-21 shows the kaleidoscope arcs resulting from pencil crystals with long axes horizontal but with all other possible orientations. These are calculated to come from exactly the same crystals, with the same orientations, as the Wegener arcs, and it may be that both sets of arcs can occur together. There have been a few reports of two sets of intersecting anthelic arcs,[10]

8 6

but photographs are clearly needed to make the comparison with the models.

The cusp-shaped arcs of Plate 3-12 look as if they might result from Tricker's kaleidoscope mechanism. The vertical angle between the arcs is too small to make them Wegener's anthelic arcs, even aside from the missing arc above the parhelic circle. Although the arcs look like the kaleidoscope arcs for a sun elevation of 30 degrees, the sun elevation as measured from the photograph, and as calculated from the time and geographical location at which it was taken, comes out to be 21 degrees. The angle between the arcs measures 50 ± 5 degrees. The kaleidoscope simulations gives an angle of 46 degrees for this elevation; so there is reasonable agreement between the two. The simulation for 20 degrees shows a loop extending above the parhelic circle. It is possible that our calculations, without any intensity factors, misrepresent the intensity of that loop. This is probably Tricker's kaleidoscope arc, matching the angular shape of the simulation; but we need to compare more photographs to be confident of the explanation.

THE ANTHELIC PILLAR AND ANTHELION

Hastings[11] described two other mechanisms that he claimed should form anthelic arcs. They are shown in perspective view and in end view (projection on the normal plane) in Figure 3-24A and B. Both consist of a ray entering a side face of a pencil crystal, reflecting internally off a side face and an end face, and exiting from a side face. Before discussing those two models, I propose a third, which may seem quite different from those two (Figure 3-24C). The path for this third model is that of the kaleidoscope rays, but with internal

Figure 3-24. Paths of three rays that produce the anthelic pillar for low sun elevations and the anthelion for higher elevations. For each ray there is a perspective view and an end view.

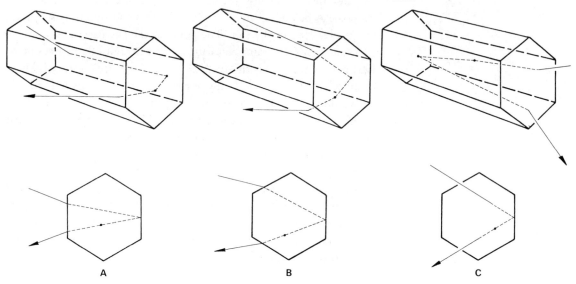

A   B   C

8 7

reflection off only one side face. One end face serves as both the entrance and exit face, and there is an internal reflection from the other end face. Although this looks quite different from the paths shown in Figure 3-24 A and B, it does essentially the same thing to the light rays.

All three of the pencil crystals in Figure 3-24 are drawn with their long axes horizontal and with a pair of side faces vertical. Though there is no evidence for such a selective orientation of pencil crystals, this orientation was used in the simulations of what we called alternate Parry arcs in Chapter 2, as a method of seeing whether there are any observations to be explained by such restricted orientations. To date we are aware of none. I will first discuss models with vertical side faces, but will then assume that the crystals can have all rotational orientations around their horizontal axes. This is the well-established distribution that gives rise to the circumscribed halo, tangent arcs to the 22-degree halo, and infralateral arcs of the 46-degree halo, all of which effects are well explained by the model simulations.

The essence of each of the drawings shown in Figure 3-24 is a ray of light being reflected from a pair of surfaces 90 degrees apart with both reflecting faces oriented vertically. Looking down from above on this double reflection we would see the diagram of Figure 3-14; the two reflections result in a deviation of 180 degrees no matter what direction the crystal axis takes in the horizontal plane. The refractions do not change that situation. They add no additional deviation in the horizontal plane; so the net result is a deviation of 180 degrees of azimuth. We should look exactly on the opposite side of the sky to see rays with this deviation. The normal-plane projections in Figure 3-24 show that, after interaction with the crystal, the ray proceeds downward, making the same angle with a horizontal plane as before. Therefore, to see the rays you would look upward at the elevation of the sun. The point in the sky described by these two conditions (the same elevation as the sun, but 180 degrees in azimuth away from it) is the anthelic point. If the ice crystals did have a pair of side faces vertical, any of the three mechanisms of Figure 3-24 would produce the anthelion, a bright spot at the anthelic point. Such an explanation has been proposed.[12] The objection to it is that none of the other effects that should result from such a distribution of ice crystals have been observed.

If we look at the light rays of Figure 3-24 for crystals taking all rotational orientations about their horizontal axes, all three should produce the same effect; the rays of Figure 3-24A and C should give the most intensity for low sun elevations, and the rays of Figure 3-24B should predominate at higher sun angles. Figure 3-25 shows the results. Instead of the predicted anthelic arcs, we are faced with

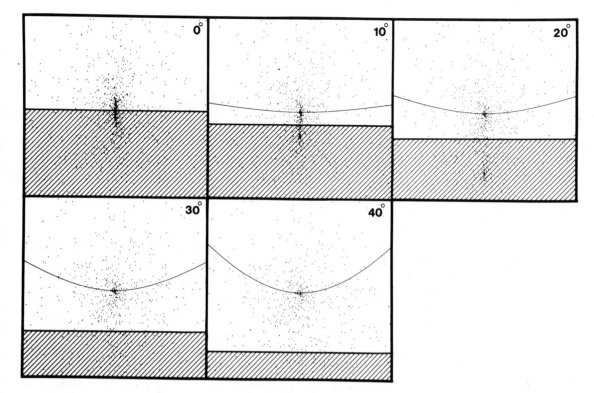

Figure 3-25. Simulations of the anthelic pillar and the anthelion resulting from the rays described in Figure 3-24.

an anthelic pillar for sun elevations from 0 degrees to about 15 degrees and with an anthelion at higher elevations! The pillar extends both above and below the anthelic point, as is the case in Plate 3-11. I determined the sun's elevation for that picture to be 6 degrees by calculation (I knew when the picture was taken) and by measurement from a photograph taken of the sun on the other side of the sky immediately afterward. You can locate the anthelic point in that photo by realizing that the shadow of my head marks the antisolar point, which is located the same distance below the horizon as is the anthelic point above the horizon.

One of the arguments in favor of these mechanisms for the anthelion and anthelic pillar is that they rely on a distribution of ice crystals that is known to exist. Lynch and Schwartz[13] have pointed out that the mechanisms of Figure 3-24A and B with a vertical pair of side faces would produce the anthelion only for sun elevations up to 46 degrees. They combed the scientific literature for all references to observations of the anthelion for which the sun elevation was given, and found that the observations existed only for sun elevations up to about 46 degrees. The same 46-degree cutoff should occur for the anthelion produced by crystals with all rotations about their horizontal axes; it does suggest that rays shown in Figure 3-24B are responsible for the anthelion at high sun elevations.

89

There is an interesting similarity between this explanation for the anthelion and the explanation that some sun pillars are caused by the external reflections from pencil crystals. In both cases the distribution of ice crystals allows light to be sent off in many directions, not restricting it to a well-defined region of the sky whose sharp boundaries define the effect. Both phenomena arise from an intensity effect — a concentration of light in a region of the sky without sharply defined boundaries. It is probably due to this circumstance that the explanation of these effects — observed since ancient times — emerged only with the computer-simulation technique.

RARE REFLECTION PHENOMENA

I have now treated most of the ice-crystal reflection and refraction effects that an alert observer is likely to see. To continue the discussion with rarer effects, which may not even have been photographed, is to risk tedium; yet there are at least two good reasons for including some of the rare effects. For one thing, having a complete list makes it possible to use the process of elimination in establishing a positive identification. It may be that we will never get the complete description of all possible halo effects, but the completeness principle is helpful in identification. Another reason for including mention of rare effects and presenting speculations on their origins is to give others the possibility of being a part of the discovery process. There is an excitement in looking for and finding something that has been seen by few people and perhaps understood by none. A good photograph may either give support to a speculative explanation or prove it to be untenable. Good photographs are needed for understanding, and in this enterprise a serious amateur is on equal ground with the scientist.

I will compromise, not treating every reliable observation that has been made, but including a few more effects that the reader may take as a challenge: A photograph of any of them would be of interest and, perhaps, helpful in understanding the effect. Because some of these effects extend over large portions of the sky, there is a special problem in describing them pictorially. After discussing this problem I will return to the rare phenomena.

THE DISTORTION OF FISHEYE LENSES

To photograph effects that extend over a large part of the sky, it is convenient to use a fisheye lens, a very wide-angle lens that crowds a 180-degree field of view into a circular picture. Relatively inexpensive attachments that have come on the market in the past few years can be added to a standard lens to convert it to a fisheye lens. All of my fisheye photograph in this book (with the exception of Plate 2-15) have been taken with such an attachment. It would seem useful, then, to set up simulations of these wide-angle sky effects to match photographs taken with a fisheye lens. The problem of trying to map a 180-degree field of view on a flat piece of film is one that cannot be solved without distortion: It is the old problem of the map maker, who is faced with the need to represent, on a flat sheet of paper, features that occur on a spherical earth. The various projections available give a reasonable representation of some parts of the earth, but always with severe distortion for others. The only question is what kind of distortion you choose.

This is also the question in the design of a fisheye lens. Some expensive lenses are advertised as being equidistant; that is, for such a lens the distance of an object measured outward from the center of the photograph is proportional to its angular displacement from the camera axis. To describe it another way: If such a fisheye lens were pointed upward at the center of the hemispherical grid structure of Figure 3-1, the north pole would be at the center of the circular photo and the equator would be the outer edge; latitude lines would appear as equally spaced concentric circles. If a feature in such a picture appears two-fifths of the way from the center to the outer edge, the object is located at a zenith angle of two-fifths of 90 degrees. When I calibrated the angular mapping of my fisheye attachment,[14] I was surprised to find that up to an angle of 80 degrees from the camera axis (which is as far as my measurements extended), the distance from the center varied linearly with angle; mine is also, then, an equidistant lens. Fisheye lenses can be made with other distortion characteristics, but all the fisheye simulations shown here use this linear, or equidistant, projection.

THE COMPLETE WEGENER ANTHELIC ARCS

I have discussed mechanisms for the anthelic arcs, some of which also produce arcs in other parts of the sky, away from the anthelic point. For example, the rays that produce the Hastings anthelic arcs (Figures 3-20 and 3-21 are rays identical to those that produce the Parry arcs, but with the addition of one internal reflection off the end face. If you think of rotating the crystal in its horizontal plane so that the ray inside the crystal just grazes the end face, the same path would contribute to the Parry arc and to the arc that, extended, crosses the anthelic point. That is to say, on the sun side of the sky the arcs should touch. This should be true for Hastings anthelic arcs touching the Parry·arcs and for the Wegener anthelic arcs touching the 22-degree halo.

Figure 3-26 shows the fisheye simulations of the Wegener anthe-

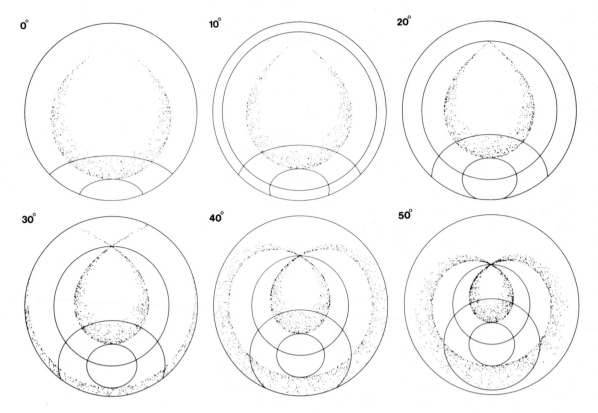

0°   10°   20°

30°   40°   50°

Figure 3-26. The simulations of the complete Wegener anthelic arcs shown in a zenith-centered, fisheye projection. The lines indicate the horizon, parhelic circle, and 22- and 46-degree halos.

lic arcs and their extensions on the sunward side of the sky. For high enough elevation of the sun, the lower arms of the anthelic arcs sweep around to touch the bottom of the 22-degree halo. Chapter 4 will show some indications of the reality of these loops toward the sun.

SUBHELIC ARCS

When you use a kaleidoscope, you observe light that comes in one end and out the other, making various numbers of reflections from the side mirrors. Let us consider rays going through a horizontal pencil crystal in this manner, making two internal reflections off the side faces (Figure 3-27). If the crystal has all rotational orientations around the horizontal axis, the result is a pair of arcs intersecting at the subsun — the subhelic arcs. The simulations of Figure 3-28 show the prediction.[15]

Figure 3-27. A ray path that contributes to the subhelic arcs.

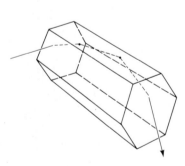

HELIAC ARCS

The previous discussion of reflection from the external faces of crystals concerned faces that are nearly horizontal (producing sun pillars

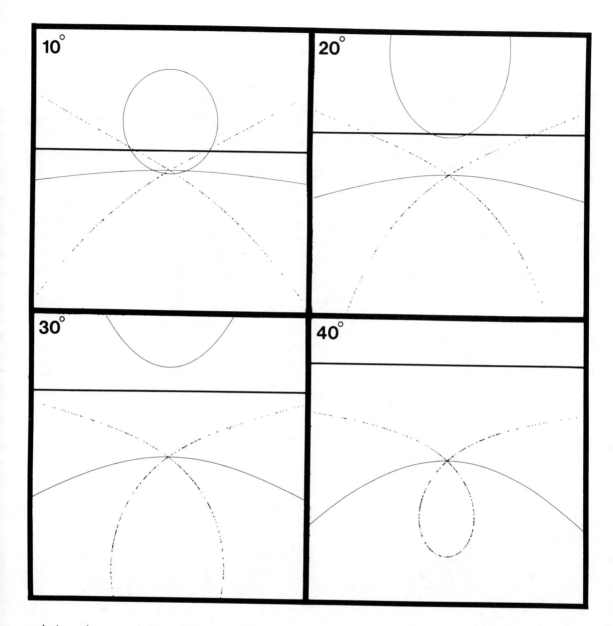

and the subsun) and faces that are nearly vertical (producing the parhelic circle). There is one more selective orientation of a crystal face that must exist for ice crystals falling through the sky. The Parry arcs give definite indication that hexagonal crystals do sometimes orient with long axes horizontal and two side faces horizontal. For such crystals the other side faces are all tipped 30 degrees from the vertical, and we might expect reflection from these faces to trace out arcs in the sky. Figure 3-29 shows the geometry of the reflection. There should be two arcs, corresponding to reflection from the upper and from the lower side faces. In general, the reflection from

Figure 3-28. Simulations of the subhelic arcs resulting from rays described by Figure 3-27. The perspective is that of a photograph centered on the subsolar point. Lines represent the 22-degree halo and the subparhelic circle.

93

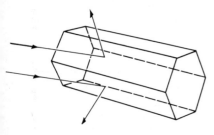

Figure 3-29. Reflections from crystal faces oriented 30 degrees from the vertical give rise to heliac arcs.

Figure 3-30. Simulations of heliac arcs shown in a fisheye projection centered on the zenith. The lines represent the horizon, parhelic circle, and 22- and 46-degree halos.

the lower face is the more likely to be visible, because you would look up (above the horizon) to see light so reflected. It is easy to show that when the reflecting face is toward the sun (i.e., when a line perpendicular to the face is pointing directly below the sun), you would see the reflected light by turning your back to the sun and looking at an angle of $(30° - \alpha)$ from the zenith, where α is the elevation of the sun. For a sun elevation greater than 30 degrees, this point moves over to the sun side of the sky. If the crystal were rotated in its horizontal plane until light from the sun just grazed the crystal face, then you would be looking directly toward the sun to see the reflected ray. That is to say, the arc should pass through the sun's position in the sky. Figure 3-30 shows the shapes of both arcs resulting from such reflections off the upper and lower side face. Reflection from the lower face produces the arcs above the sun, and reflection from the upper face produces the continuation of these arcs below the sun.

A photograph by Austin Hogan (Plate 4-2) shows, among many other interesting effects, a faint arc concave toward the zenith opposite the circumzenithal arc. From the photo I measured the distance from the zenith as 13 degrees and the sun elevation as 18 degrees, which should give a zenith distance of 12 degrees $(30 - 18)$. The agreement was surprising to me and quite exciting;

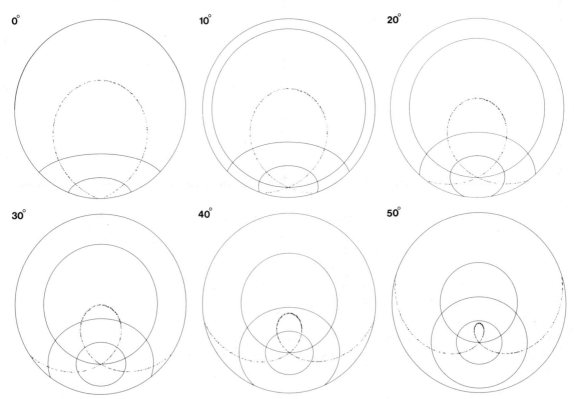

9 4

I am not aware of any other photographs of that effect. Note that there appears to be a strong Parry arc in that display, as would be expected if our explanation is correct. Humphreys[16] has discussed these arcs and referred to them as the oblique heliac arcs. I suggest that the name *heliac arcs* is sufficiently descriptive to identify this effect.

SUBANTHELIC (OR ANTISOLAR) ARCS

I have described the rays producing the heliac arcs as being externally reflected. Some contribution to the intensity of the heliac arcs would be made by rays that enter a pencil-crystal end face, reflect internally off one of the side faces, and exit from the other end face.

Ice-crystal reflection effects

Figure 3-31. Simulations of the subanthelic (or antisolar) arcs centered on the antisolar point. Lines represent the horizon and the subparhelic circles.

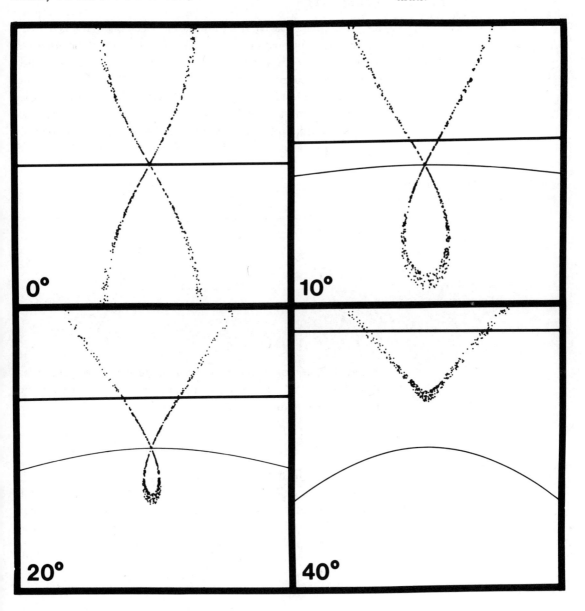

0°

10°

20°

40°

Consider now the ray that, instead of leaving the second end face, is reflected back to the end face by which it entered the crystal.[17] The additional reflection from the vertical end face should produce a pair of arcs that, from an airplane, could be seen intersecting at the antisolar point. I call them subanthelic arcs, because they cross at the subanthelic point and have the form of Tricker's anthelic arcs reflected below the horizon, but they could equally well be referred to as antisolar arcs. Figure 3-31 shows the simulations with the subparhelic circle and horizon line drawn in. To decide whether these arcs can actually correspond to reality, look at the rare photograph of Plate 3-13. Subanthelic arcs, amazingly, appear to be real.

KERN'S ARC

Kern's arc, which as far as I know has not been photographed, is described as a continuation of the circumzenithal arc on the other side of the zenith. Its explanation would appear to require internal reflection from the side face of a pencil crystal and subsequent emergence from a different side face. The crystal, according to most proposed explanations, must be fairly long compared with its width, but such a crystal would not be expected to have the appropriate orientation. Figure 3-32 shows a possibility suggested by Tricker;[18] the result

Figure 3-32. Rays suggested to produce the circumzenithal arc and Kern's arc.

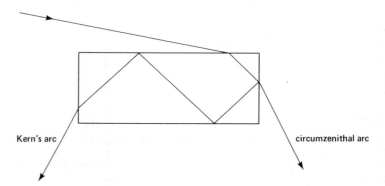

should be quite faint, however, because three internal reflections, none of them total, are involved. A photograph would give us something to work with.

HEVEL'S HALO AND THE 90-DEGREE PARHELIA

The literature includes some references to a halo, centered on the sun, with an angular radius somewhere around 90 degrees. Usually what is reported is some small portion of it, crossing the parhelic circle about 90 degrees from the sun. Humphreys[19] suggests that the 90-degree parhelia are just the bright spots resulting from the

intersection of the parhelic circle with the 90-degree halo, but, as
we will see, this is not a tenable explanation. Part of the difficulty in
explaining these features is the lack of reliable observations. The
most detailed description of which I am aware is an early drawing by
Hevelius,[20] a Danish astronomer, made in the city of Gdansk on the
Baltic Sea in 1661. His drawing is reproduced and discussed in
Chapter 4 (Figure 4-1), and from his report the halo is frequently referred to as the halo of Hevelius or Hevel's halo. The angular radius
of such a large halo is difficult for even a scientifically sophisticated
observer to estimate. Note that the angular distance around the
parhelic circle (the azimuthal angle) does not give the correct radius
unless the sun is on the horizon. The angular distance must be
measured along a great circle. For example, if we refer to a parhelion
of 90 degrees, we refer to a spot on the parhelic circle, halfway between the sun and the anthelic point; that is, at an *azimuthal* angle
of 90 degrees from the sun. The *actual* angular distance from the sun
is less than 90 degrees. As the sun rises higher and higher in the sky,
its angular distance from the 90-degree parhelion becomes less and
less.

It has been suggested that Hevel's halo comes from light taking
the path shown in Figure 3-33. In this mechanism, light enters a
side face of the hexagonal crystal, reflects off two alternate side faces,
and emerges from the entrance face. The total deviation of such a ray
results from the deviation of 240 degrees contributed by the two
reflections, plus an additional 22 degrees of minimum deviation
contributed by the two refractions. With a total minimum deviation
of 262 degrees, one would look in a direction 98 degrees from the
sun (360 − 262) to see the minimum-deviation edge of the halo; but
in this unusual situation, this sharp edge should be on the side of the
halo away from the sun. Figure 3-34 shows the simulation for rays
taking such paths through crystals with random orientations. Remember that when we are dealing with reflections we have no intensity factors in the calculations. Such factors would cause the
simulated intensity to fall off more rapidly away from the
minimum-deviation edge, and the result would probably look more
like a halo. However, what is obvious is that this halo of 98-degree
radius, for a sun elevation of 26 degrees, does not intersect the
parhelic circle anywhere near the positions of 90-degree parhelia.

Suppose we consider plate crystals with their bases nearly horizontal − the kind of distribution that produces the sun dogs (22-degree
parhelia). Figure 3-35 shows the resulting pattern for the rays (of
Figure 3-33) going through such a distribution, along with the
sun dogs produced by the same set of crystals. This pattern is most
interesting, because the concentrations of light are centered very
near the 90-degree-parhelion points. The angular position from the

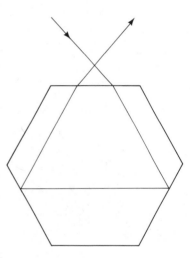

Figure 3-33. Ray path that could
produce 90-degree parhelia.

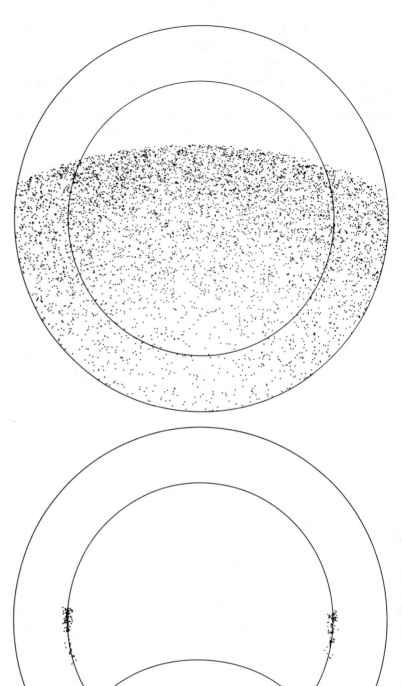

Figure 3-34. The effect resulting from the rays defined in Figure 3-33 passing through randomly oriented plate crystals for a sun elevation of 26 degrees.

Figure 3-35. Simulation of 22-degree parhelia and 90-degree parhelia from light passing through plate crystals with nearly horizontal bases. The 90-degree parhelia result from rays described by Figure 3-33 at a sun elevation of 26 degrees.

sun is less than 98 degrees because the minimum deviation for
skew rays is less than the minimum deviation for normal rays passing through the crystal. (To understand this point it may help to
review the discussion of skew rays and normal rays in the section of
Chapter 2 dealing with 22-degree parhelia.) The crystals for this
simulation have a maximum-tilt angle of 2 degrees. If we let them
tilt more, the 90-degree parhelia spread out in all directions but do
not develop arcs.

To conclude, then: With this mechanism we can produce 90-
degree parhelia or a somewhat larger halo (with an outside edge of
radius 98 degrees), but we cannot produce a halo passing through
the 90-degree parhelia as in the drawing of Hevelius.

I have one more idea for explaining the halo of Hevelius. In Figure 3-28 the subhelic arcs are shown crossing the subsun point. The
upper branches rise above the horizon and form a loop around on
the other side of the sky. For a sun elevation of 26 degrees — the estimate for Hevelius's drawing — the subhelic arcs cross the parhelic
circle, not at 90 degrees, but at 115 degrees. However, the azimuth
of the intersection varies rapidly with sun elevation, and at a sun
elevation of only 3 degrees less (23 degrees), the subhelic arc crosses
the parhelic circle near 90 degrees, as shown in the simulation of
Figure 3-36. This arc is not exactly a circle around the sun, but it is

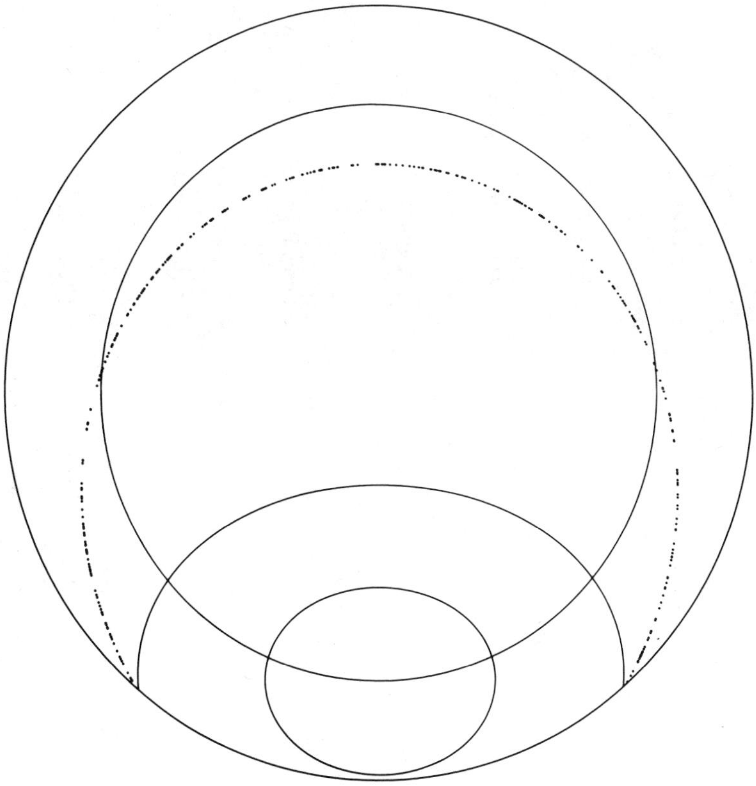

Figure 3-36. Simulation of the subhelic arc for a sun elevation of 23
degrees. This may be a candidate for
Hevel's halo.

99

Table 3-1. *Effects produced by reflection and refraction from ice crystals, classified by crystal type and orientation*

Crystal type and orientation	Name of effect	Path of light ray	Figure showing simulation of effect	Comment
Pencil or plate crystals with random orientations	22° halo	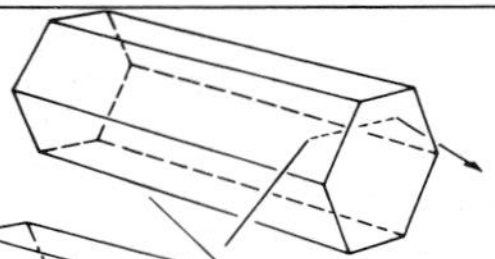	2-7	Seen for all sun elevations
	46° halo	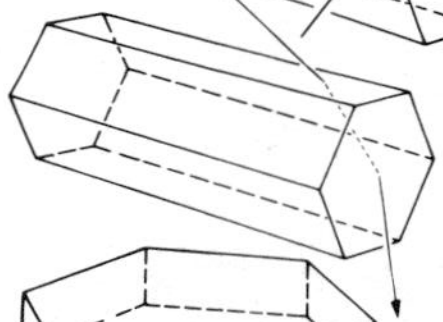	2-25	Seldom, if ever, seen complete
Plate crystals with nearly horizontal bases	22° parhelia (sun dogs)	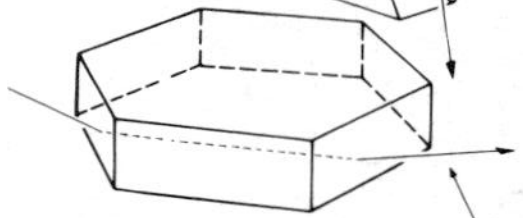	2-8	Cannot be formed for sun elevations greater than 61°; additional internal reflection off base results in subsun dogs
	120° parhelia	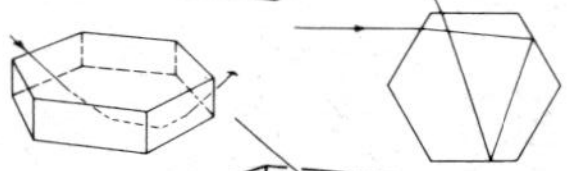		Probable mechanisms for 120° parhelia: sketch on left for high sun elevations, on the right for low elevations
	Circumzenithal arc	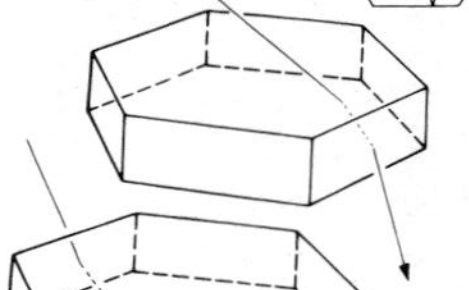	2-28	Can occur only for sun elevations below 32°
	Circumhorizontal arc	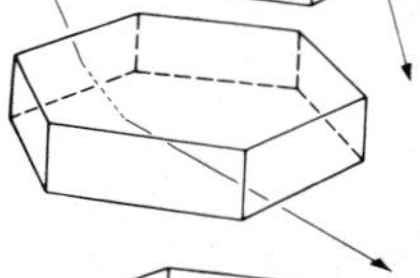	2-29	Can occur only for sun elevations greater than 58°
	Sun pillar and subsun	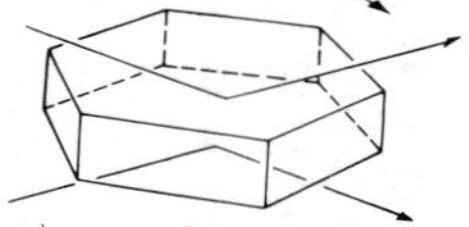	3-4	Can also be produced by reflection from sides of pencil crystals with long axes horizontal
	Parhelic circle	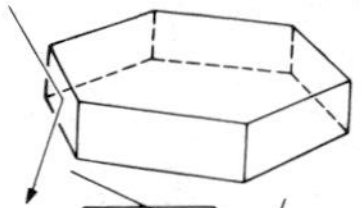		Can also be produced by reflection from ends of pencil crystals with long axes horizontal
Plate crystals spinning about a horizontal axis that passes through opposite points of a hexagonal base	Lowitz arcs	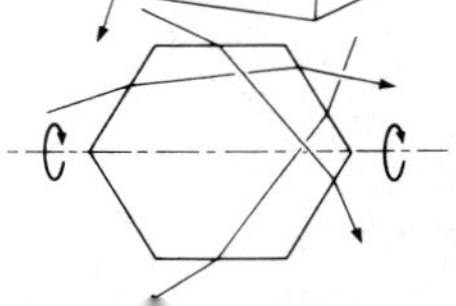	2-23	Explanation not well verified; few photos available

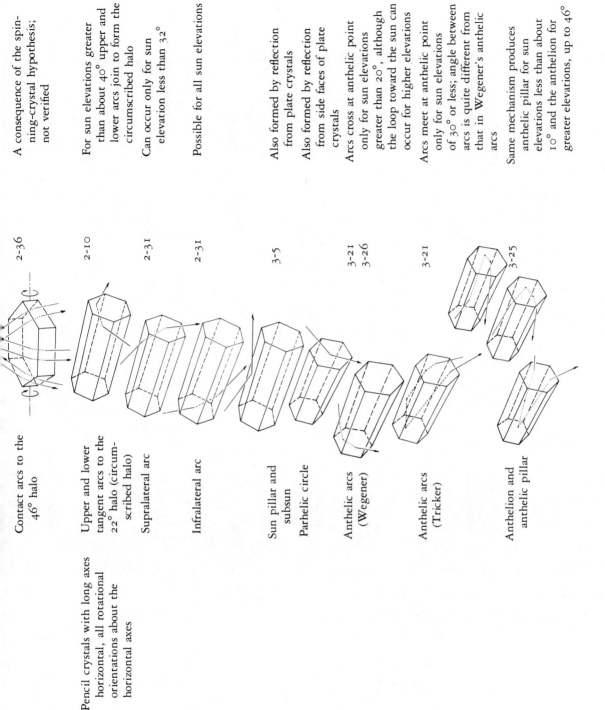

Pencil crystals with long axes horizontal, all rotational orientations about the horizontal axes

Name	Fig.	Description
Contact arcs to the 46° halo	2-36	A consequence of the spinning-crystal hypothesis; not verified
Upper and lower tangent arcs to the 22° halo (circumscribed halo)	2-10	For sun elevations greater than about 40° upper and lower arcs join to form the circumscribed halo
Supralateral arc	2-31	Can occur only for sun elevation less than 32°
Infralateral arc	2-31	Possible for all sun elevations
Sun pillar and subsun	3-5	Also formed by reflection from plate crystals
Parhelic circle		Also formed by reflection from side faces of plate crystals
Anthelic arcs (Wegener)	3-21 3-26	Arcs cross at anthelic point only for sun elevations greater than 20°, although the loop toward the sun can occur for higher elevations
Anthelic arcs (Tricker)	3-21	Arcs meet at anthelic point only for sun elevations of 30° or less; angle between arcs is quite different from that in Wegener's anthelic arcs
Anthelion and anthelic pillar	3-25	Same mechanism produces anthelic pillar for sun elevations less than about 10° and the anthelion for greater elevations, up to 46°

Table 3-1. (*cont.*)

Crystal type and orientation	Name of effect	Path of light ray	Figure showing simulation of effect	Comment
	Subhelic arcs		3-28	Can also be produced by pencil crystals with pairs of faces horizontal
Pencil crystals with long axes horizontal and one pair of side faces oriented horizontally	Parry arcs		2-10	Upper sunvex arc can be seen for sun elevations up to about 15°; upper suncave arc for sun above 5°; lower sunvex arc for sun elevation up to 50°; lower suncave arc for sun above about 40°
	Anthelic arcs (Hastings)		3-21	Arcs cross at anthelic point only for sun elevations greater than about 30°, although the sunward loop can occur for higher elevations
	Heliac arcs	Plus other internal reflection mechanisms	3-30	Possible for any sun elevation
	Subhelic arcs		3-28	Same as for pencil crystals with long axes horizontal without the additional Parry-arc-orientation restriction
	Subanthelic arcs		3-31	Arcs cross at subanthelic (antisolar) point only for sun elevations less than 30°

difficult to judge the shape of such a large-scale phenomenon and it might well be described as a circle. Perhaps there are few observations of the arcs intersecting the parhelic circle at 90 degrees because this condition can occur only with a sun elevation near 23 degrees.

I have suggested some possible explanations for Hevel's halo. To evaluate them we need a new observation to supplement the original one reported over three centuries ago. The next time I hope we have a photograph.

SUMMARY OF ICE-CRYSTAL EFFECTS

I have treated many effects in these two chapters — far more than most people can absorb in one reading. I have classified them as refraction or reflection phenomena and have further divided the discussion into 60-degree- and 90-degree-prism effects. Another possible classification is to consider together all of the effects that can come from crystals of a given shape with a particular distribution of orientations. Table 3-1 follows this approach, using five different classifications of crystal type and orientation. A drawing indicates the ray path through the crystal, and the figure number of the resulting simulation is given for easy reference.

Figure 3-37. Puzzle photograph, explained in the Appendix. (Courtesy of the Flandrau Planetarium, University of Arizona, Tucson)

PUZZLES INVOLVING REFLECTION

I give you here three puzzles, two for which I have answers and one for which I do not.

Figure 3-37 is a fisheye photograph taken at night. The moon is the source of illumination. What can you see (and understand) – before checking the Appendix?

Plate 3-14 is a view from an airplane showing several interesting optical effects below the horizon, all of which I have discussed. What do you see?

I have no answer in the Appendix for the photograph of Plate 3-15. The rings surrounding the subsun spot are called Bottlinger's rings, and I do not understand their formation. If they were diffraction rings resulting from the small reflecting ice plates, I would expect to see them colored like the coronas of Chapter 6. One of the more plausible explanations is that they are produced by multiple reflections between flat-plate ice crystals, each with a small tilt from the horizontal. Thus the subsun results from one reflection, the first ring from three reflections, the second from five, and so on. It seems that the form of the rings might be explained this way; yet I suspect that such a mechanism would produce too low an intensity in the rings to be observable. The question is yet to be answered.

4

Complex displays, past and present

The development of the computer simulations of Chapters 2 and 3, and the process of coming to understand them, took place over a period of a dozen years. When my colleagues and I first studied the shape of a sun pillar, it was certainly not clear where our effort would lead. Finally we realized that we had treated, one by one, most of the commonly observed effects and that we should have all of the pieces necessary to understand something as complex as the marvelous St. Petersburg display of 1790.[1] So we instructed the computer to produce a master simulation made up of a basic 22-degree halo, along with a generous circumscribed halo and a good dose of infralateral arcs, seasoned with a dash of sun dog, garnished with a parhelic circle – following Tobias Lowitz's old recipe for a St. Petersburg Delight.

When we got the result, there was real excitement to see whether, by trying to understand the elements, piece by piece, we had managed to gain insight into the whole. In fact, we had. At first glance, it appeared that we had matched the old drawing; at second glance we could see differences, some of which revealed things we had not understood from the separate pieces; we also could conclude some interesting things about how Lowitz had made his observations. At this point a colleague who was not directly involved in the project stepped into the office and, in response to the excitement, gave us a name for our enterprise: "sky archeology."

Of course, the antiques we are dealing with are the records, not the effects. We assume that the same kind of ice crystals existed in 1790 as today and that the laws of reflection and refraction have not changed. It has been suggested that Ezekiel's vision of wheels within wheels may be an earlier description of a complex display of ice-crystal phenomena. I assume that the sky displays we see these days have existed in ages past, even predating human vision to perceive them.

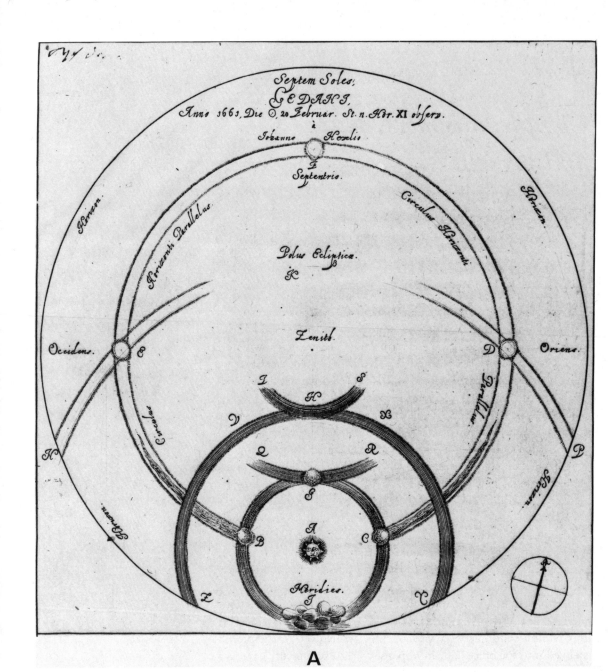

Septem Soles
GEDANI
Anno 1661, Die ☉ 20 Februar. St. n. Hor. XI observ.
à
Ioanne Hevelio

A

Figure 4-1. A: Hevelius's "Seven
Suns" drawing of 1662. B: Simulation of Hevelius's drawing. (A, from
Johannes Hevelius. *Mercurius in sole
visus* . . . S. Reininger, 1662)

HEVEL'S GDANSK (OR DANZIG) PHENOMENA

Let us begin with an earlier observation than that of Lowitz. The
Danish astronomer Hevelius[2] (or Hevel) produced a drawing of a
display he observed in Gdansk on February 20, 1661 (Figure 4-1A).
(The city was named Gdansk when Hevelius made his observation,
as it is today. Because its name has been Danzig in intervening
years, the display has also been referred to as the Danzig Phenom-

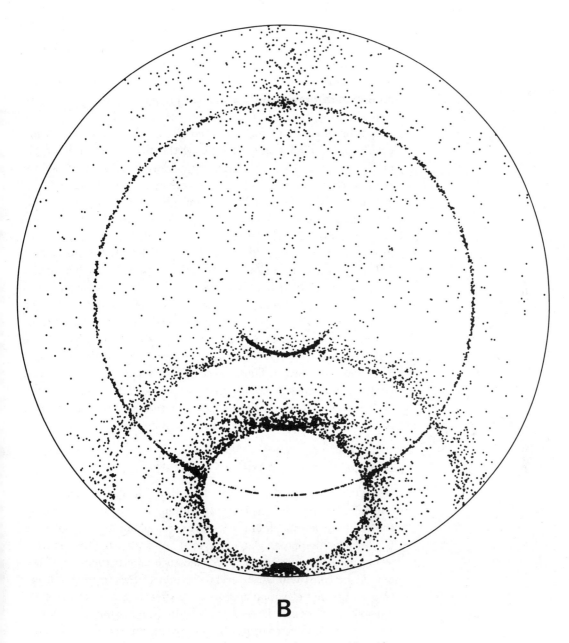

B

ena.) The title of the drawing is "Seven Suns," and you can identify the suns as the features represented by small circles.

The sun appears to be about 26 degrees above the horizon, and many of the elements of the drawing are familiar to us: the 22- and 46-degree halos, 22-degree parhelia, upper tangent arc to the 22-degree halo, circumzenithal arc, and parhelic circle. Hevelius shows as one of his suns the anthelion, which is surprising, but most unusual are the arcs crossing the parhelic circle about 90 degrees from the sun. These arcs are generally assumed to be part of a 90-degree halo that, from this report, is sometimes called Hevel's halo.

The drawing of Hevelius shows the entire sky projected within a circle representing the horizon. One of our decisions was what kind of mapping we should choose for the simulation of the drawing. Most drawings show each of the halos as a circle. In the equidistant projection that my fisheye camera lens produces, however, the halos near the edge of the picture are squashed vertically. There is one projection in which all circles plotted on the celestial sphere will end up mapped as circles on the flat plane: the map makers' isomorphic projection. But although concentric circles on the celestial sphere get plotted as circles using this projection, the circles are not concentric on the plane projection. I am quite certain that Hevelius's drawing was made intuitively, not produced according to any specific projection. For the simulations we decided to choose the equidistant projection and suffer the indignity of slightly flattened halos.

For the attempt to simulate Hevelius's "Seven Suns" we used a random crystal distribution to produce the halos; pencil crystals with long axes horizontal to produce the upper tangent arc, the anthelion, and, by reflection from the end faces with up to 0.5-degree tilt, the parhelic circle; and plate crystals with tilts up to 1 and 2 degrees to produce the circumzenithal arc and the sun dogs. The result is shown in Figure 4-1B. Hevel's halo was not included because, as discussed in Chapter 3, we do not fully understand it; however, over three hundred years after the drawing was made, I think we understand the other features quite well.

LOWITZ'S ST. PETERSBURG DISPLAY

Let us take a look at the entire St. Petersburg display (Figure 4-2A). Lowitz[3] describes the display as beginning at 7:30 in the morning and reaching its maximum development at about 10:00, with the last features disappearing at 12:30 P.M. At 10:00 the sun elevation would have been near 50 degrees, and such an elevation would appear to be consistent with several features of the drawing; for example, the sun's elevation appears to be slightly greater than the radius of the 46-degree halo. Figure 4-2B shows a simulation for an elevation of 50 degrees using the equidistant projection. It includes the results of random crystal orientations to give the two halos; horizontal pencil crystals to give the circumscribed halo, infralateral arcs, anthelic arcs, and (with 0.5-degree tilt of the axis) the parhelic circle; plate crystals with nearly horizontal bases (± 2 degrees) to add 22-degree parhelia and 120-degree parhelia; and plate crystals spinning about a horizontal axis to add the arcs that have acquired Lowitz's name.

Lowitz described the 22-degree halo and circumscribed halo as two intersecting circles with their centers at α and β (displaced horizontally). They are matched, however, quite well by the simulation. As I commented in Chapter 3, the upper Lowitz arcs are partially obscured by their approximate coincidence with the circumscribed halo for this sun elevation; this perhaps explains why they were not recorded on the drawing. The sunward loop of the anthelic arcs is more rounded in our simulation than in the drawing, but its general shape is otherwise quite satisfactory. The shape of the infralateral arcs in the flat-plane projection of Figure 2-24B matches Lowitz's drawing better than the fisheye mapping of Figure 4-2B. I suspect that this is because Lowitz first drew the effects near the sun using the same perspective as in a conventional photograph (including a flat horizon) and then added the other effects to the same drawing. For a sun elevation of 50 degrees, the parhelic circle should have a radius of 40 degrees, less than that of the 46-degree halo, although Lowitz shows it as much larger. But these are just mapping problems, expressive of the difficulty of representing the hemisphere of the sky on a flat sheet of paper. There are more significant discrepancies.

Two things are missing in the simulation: the arc tangent to the top of the 46-degree halo, and the arc tangent to the bottom of the 22-degree halo. The arc at the top of the large halo looks exactly like a circumzenithal arc: It is a circular arc, centered on the zenith. But this arc cannot appear, according to our theory, when the sun is above a 32-degree elevation. The arc at the bottom of the halo is also a puzzle. We have no predictions nor any other reports of such a feature, but I think that we can find the clue to the solution of both these problems in the drawing itself.

Several elements of the drawing are consistent with a sun elevation of 50 degrees; at such an elevation the circumscribed halo has the form shown, and the bottom of the 46-degree halo lies just above the horizon, as indicated. However, with the sun at a 50-degree elevation, the top of the 46-degree halo would extend beyond the zenith. Stated another way, the 46-degree halo (with a diameter of over 90 degrees) cannot fit between the zenith and horizon with space to spare, as shown in the drawing.

When the sun is lower in the sky, both of the missing arcs could appear. At an elevation around 25 or 30 degrees the lower tangent arc or the lower sunvex Parry arc (Figure 2-10) would appear approximately as shown, and the circumzenithal arc could circle the zenith as in the drawing. It appears that Lowitz's drawing is not a "snapshot picture" of the display as it appeared at 10:00 A.M. but rather a composite of sketches he made over the long duration of this spectacular display.

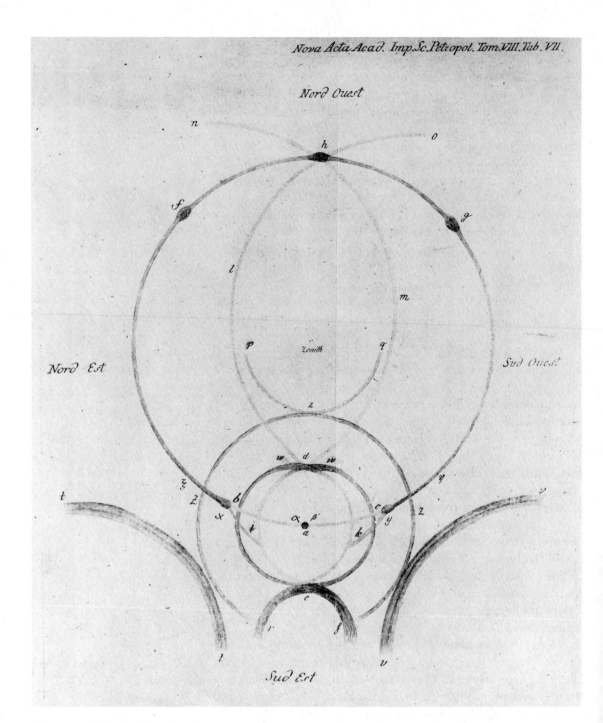

Figure 4-2. A: Lowitz's drawing of
the St. Petersburg display of 1790.
B: Simulation of Lowitz's drawing.

A

PARRY'S OBSERVATION

It was about thirty years later that Parry[4] produced his sketch showing the arc now commonly known by his name. Figure 4-3 shows
the sketch and the simulation made with the perspective of a camera

110

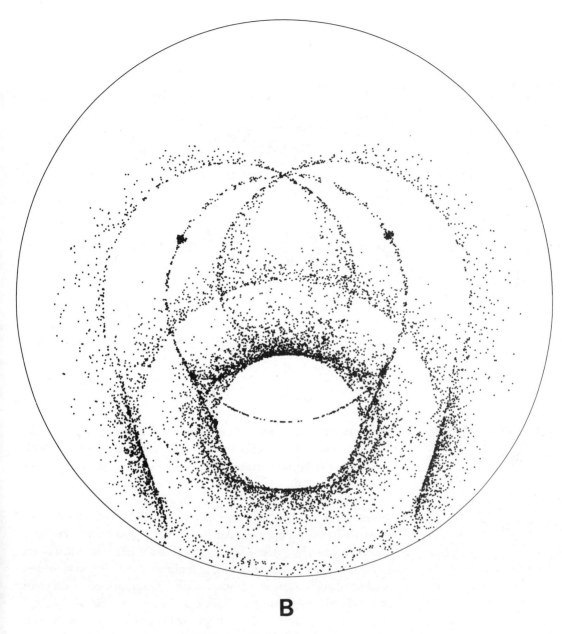

B

pointed at the sun with the sun 22 degrees above the horizon. The upper suncave Parry arc is well represented by the simulation. The clear presence of the infralateral arcs for the sun elevation of 23 degrees implies that the supralateral arc should also be present. In the simulation the supralateral arc adds intensity to the top of the 46-degree halo, in agreement with Parry's use of solid lines for that part of the circle, rather than the dotted lines he used for other parts. The lower tangent arc, mostly cut off by the horizon, accounts for the brightness at the horizon in Parry's sketch. The circumzenithal arc could be formed by the plate crystals that are responsible for the

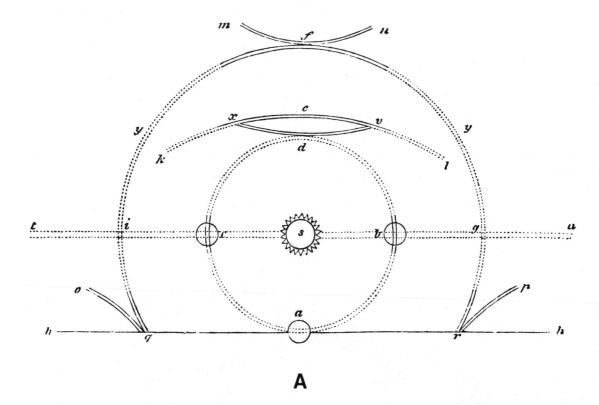

A

Figure 4-3. A: Parry's drawing of his observation of 1820. B: Simulation of Parry's drawing.

sun dogs or could result from light passing through the doubly oriented pencil crystals that produce the Parry arc. The responsible rays enter the upper horizontal face and exit from the vertical end face of the crystals.

Two small discrepancies remain. Parry's use of a straight line to represent the parhelic circle is not surprising, given the need to show the relationships in a flat plane. Parry also shows the infralateral arcs touching the 46-degree halo at the horizon, whereas the simulation shows them touching at a higher point. If you have ever tried to sketch such phenomena, with the intensities of different parts waxing and waning and never as distinct as they end up appearing on your sketch, I think you will agree that the agreement between sketch and simulation is quite good.

BAVENDICK'S ELLENDALE DISPLAY

We now skip one hundred years to a display in Ellendale, North Dakota, recorded as a drawing by Frank Bavendick[5] on March 8, 1920. This hundred-year gap between the records should not be interpreted as a measure of the frequency of occurrence of such grand displays: Other records have been made and reported. The rarest feature is probably the presence of an observer who is skilled enough to

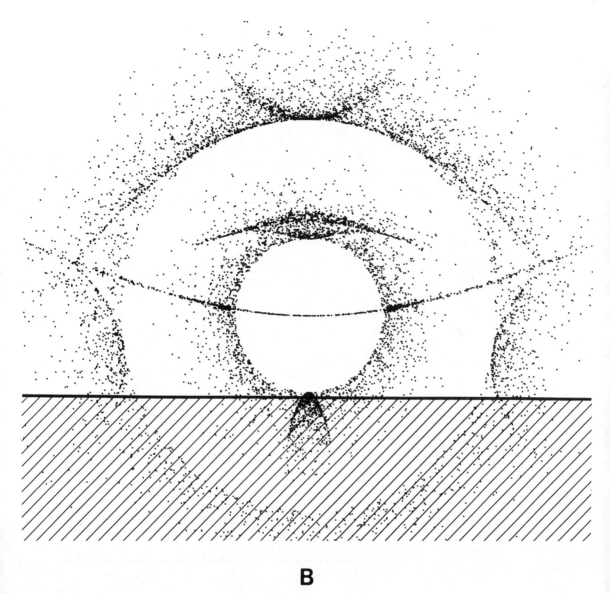

B

record what he or she sees with reasonable objectivity and who is comfortable enough with the scientific enterprise to publish the description. The present-day proliferation of good cameras, which give interested amateurs the capacity for objective recording of events, should greatly increase the possibility of good records of such spectacular displays.

Figure 4-4A shows Bavendick's drawing of the Ellendale display. The mapping, with the straight horizon, is much like that in Lowitz's drawing. From the drawing the sun elevation appears to be about 38 degrees. This agrees with the calculated elevation of the sun at 1:30 P.M., when the display was said to be as shown in the

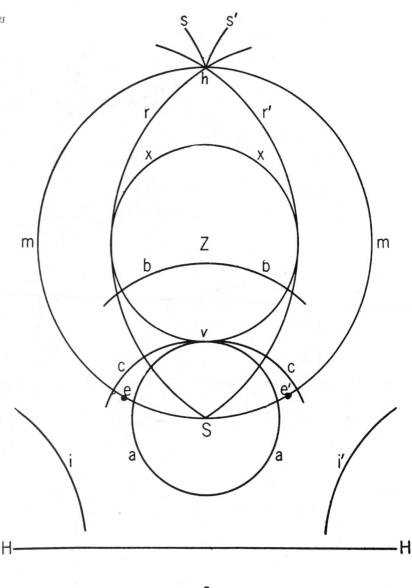

A

Figure 4-4. A: Bavendick's Ellendale display of 1920. B: Simulation of the effects in Bavendick's drawing for a sun elevation of 38 degrees. C: Simulation of Bavendick's drawing for a sun elevation of 30 degrees.

drawing. The 22-degree halo, sun dogs, infralateral arcs, and parhelic circle are all rather obvious by now. Arc cc looks like the top of the circumscribed halo (shown in the simulation of Figure 4-4B), though it is a bit surprising that the lower part is entirely missing. The supralateral arc cannot appear for a sun elevation this high, and so arc bb would appear to be the upper part of the 46-degree halo, with the rest missing. The circle centered on the zenith is not the circumzenithal arc, because it is shown tangent to the 22-degree halo. It presents an interesting problem. Bavendick suggests it might be a "secondary parhelic circle" resulting from the great in-

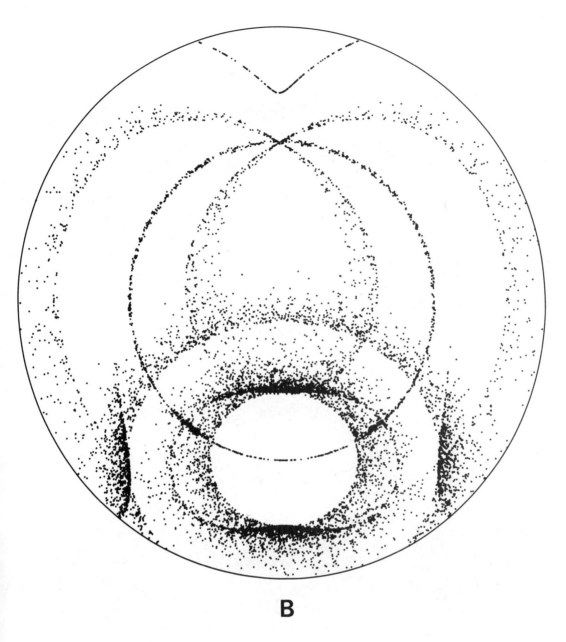

B

tensity at the top of the 22-degree halo. My guess is that it results from the overlapping of two different effects, only one of which I understand. The anthelic arcs that loop around the sky to touch the top of the 22-degree halo (as shown in Figure 3-26) seem to match the drawing very well. To complete the drawing we need another arc that passes through the sun and loops around the other side of the zenith, as does arc xx. The heliac arcs (Figure 3-30) come to mind. They do have this shape for low sun elevation, but for an elevation of 38 degrees the heliac arc does not reach as high as the zenith. None of the effects we have treated till now have this shape; there is a

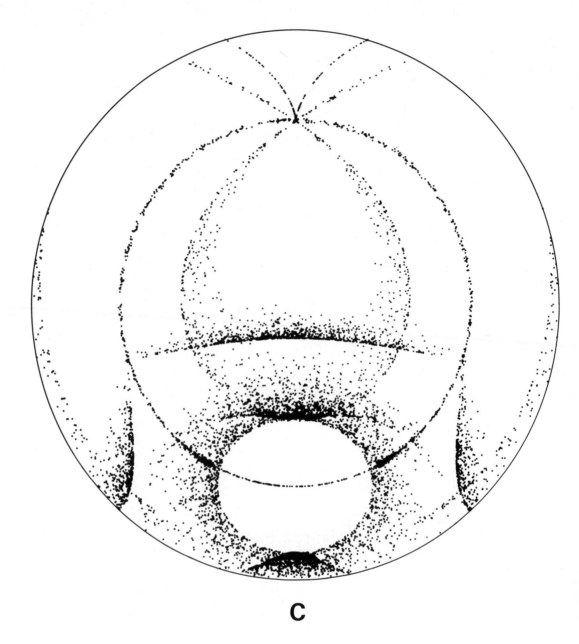

C

problem yet to be solved. The two sets of anthelic arcs appear to present no problem at first glance, the one set coming from Wegener's mechanism and the other set (marked SS′ in the sketch) having the shape predicted by Tricker for his kaleidoscope mechanism and illustrated by Plate 3-12 and Figure 3-21. However, for sun elevations above 30 degrees, the kaleidoscope arcs do not intersect the anthelic circle.

The simulation in Figure 4-4B includes the effects that best seem to match the drawing for a sun elevation of 38 degrees. The overall representation of the drawing is poor. Figure 4-4C, however, shows a simulation for a sun elevation of 30 degrees and, with the exception of the 22-degree halo and the sun dogs, includes only effects that can come from pencil crystals with long axes horizontal. This simulation includes the circumscribed halo, the infralateral and supralateral arcs, the parhelic circle, Wegener's anthelic arcs, and the kaleidoscope anthelic arcs of Tricker. The arcs through the anthelic point look very much like those of the drawing. The supralateral arc seems to be a better match for arc bb than the 46-degree halo of the simulation in Figure 4-4B. The one thing missing is an arc passing through the sun and continuing around the zenith, to match arc xx of the drawing. The heliac arc (see Figure 3-30) has the right general shape for a sun elevation of 30 degrees but passes through the zenith — not a likely explanation for the observation. I would say that the arc labeled xx represents an unsolved problem, but the agreement of all the other features for simulations over the range of sun elevations from 30 to 38 degrees looks good. It is at least impressive enough to make me wonder, again, whether over the time taken to make the drawing and to make some angular measurements with weather balloon theodalites (mentioned in the original report) the sun's elevation could have changed significantly. In the absence of a photograph, we will not know for sure.

BLAKE'S ANTARCTIC OBSERVATION

J. R. Blake[6] spent some time on the Antarctic ice cap in the austral summer of 1958–9 and reported many intriguing halo effects, some of them well described for the first time. The drawing of Figure 4-5A is, perhaps, the most spectacular of his recorded observations. The 22- and 46-degree halos are familiar, along with the sun dogs, parhelic circle, and circumzenithal arc. From the shape and extent of the sun pillar, I conclude that it is a result of reflection from pencil crystals with long axes horizontal. Just above the 22-degree halo is, not one arc, but three. The lowest is the upper tangent arc; the next is the upper sunvex Parry arc; and the highest of the three, the upper suncave Parry arc. The sun elevation lies in that rather narrow, 10-to 15-degree range where both of the upper Parry arcs can be seen together. In fact, the relative positions of the two Parry arcs are a very sensitive function of sun elevation, and the simulation shown in Figure 4-5B for an elevation of 11 degrees reproduces their position considerably better than one for the 13 degrees that Blake reports for the sun elevation at the time of the observation. The remaining effects are less familiar.

The anthelic pillar is not common, but I have photographed it (Plate 3-11) and we have simulated it (Figure 3-25). We can also add a sun pillar resulting from external reflection off the sides of pencil crystals. The four features that Blake has numbered in his

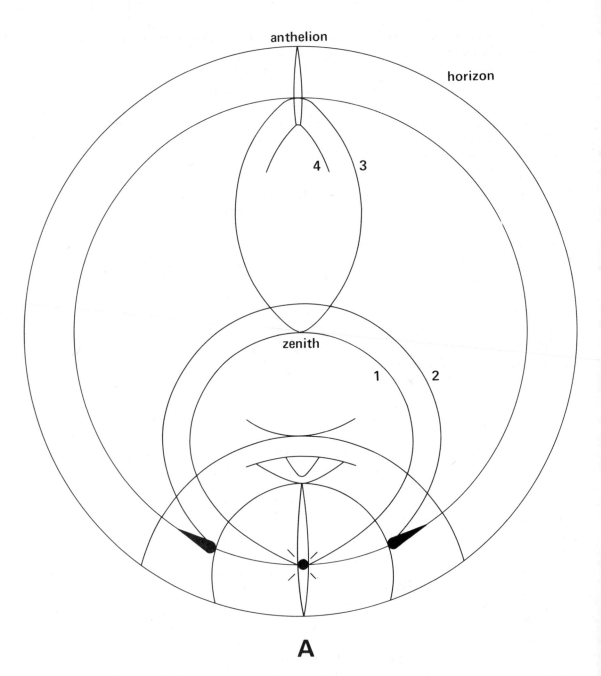

anthelion

horizon

4 3

zenith

1 2

A

Figure 4-5. A: Blake's Antarctic observation of 1958. B: Simulation of Blake's drawing.

drawing are still to be explained. The presence of the strong Parry arcs indicates the existence of some pencil crystals, oriented with a pair of side faces horizontal. Reflection from the other side faces of these crystals gives rise to the heliac arcs that look very much like curve 1. The heliac arc is included in the simulation of Figure 4-5B. Curve 1 appears to go through the zenith in Blake's drawing, whereas the theory of the heliac arc predicts that it should encircle the zenith, passing 17 degrees away. Blake observed this effect while

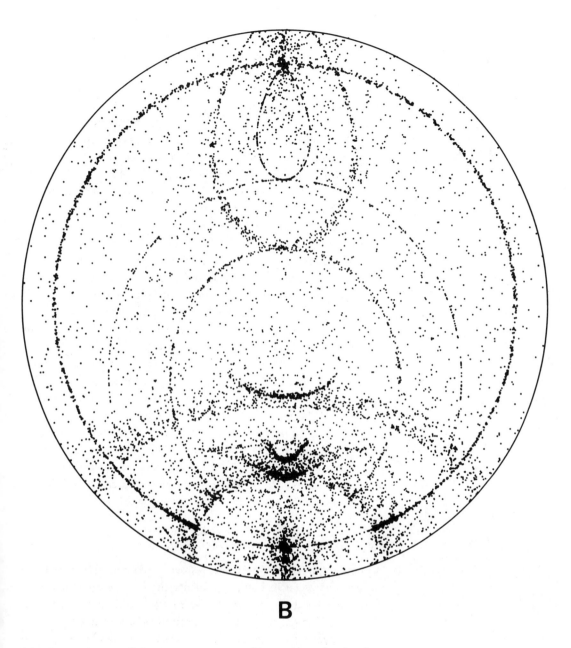

B

his Antarctic expedition party was traveling across the ice in a
tracked snow vehicle called a weasel; he comments on the 17-degree
disparity and concludes: "Although this is somewhat different from
the zenith point, it must be remembered that the position of the
celestial zenith is very difficult to estimate without some form of in-
strument, especially from a bouncing Weasel!"

What about arc 2 in Blake's drawing? There is another arc that
generally surrounds the heliac arc (1). It results from light going in
one end of a horizontal pencil crystal and out the other end, making
internal reflections from two side faces en route. The same arcs are

119

produced whether the crystal has all rotational orientations about its horizontal axis or has a pair of side faces horizontal. These are the subhelic arcs that cross at the subsolar point. Their appearance in the vicinity of the subsolar point is illustrated in Figure 3-28. The simulation of this subhelic arc, added to Figure 4-5B, looks very much like curve 2.

Arc 3 touches arc 1 and bends around to pass through the anthelic point. There is an arc that touches the heliac arc: It is the arc resulting from light that is reflected once from a side face of crystals with the Parry-arc distribution of orientations (as in the heliac arc) *and* reflected once from an end face. For the heliac arc, the single reflection can be either internal or external, but for this arc, requiring two reflections, both must be internal. These reflections produce the subanthelic arcs, whose forms near the antisolar point are shown in Figure 3-31. This arc, added to the simulation, does not pass through the anthelion, as shown in Blake's drawing, though its form elsewhere is a reasonable match for the drawing.

The only candidate I have for arc 4 of Blake's drawing is the anthelic arcs of Tricker, which in the simulation of Figure 3-21 extend farther toward the zenith, forming a loop that touches the subsun arc. The appearance of the photograph in Plate 3-12 makes me suspect that the intensity of the upper loop of our simulation may be exaggerated (because of our neglect of intensity factors). The arcs also pass through the anthelic point rather than the top of the pillar, as shown in the drawing. Taken as a whole, the simulation is not a bad match for the drawing,[7] although there are discrepancies that are shortcomings either of the theory or of the observation. Without a photograph, we cannot know for certain which is the case.

TWO SOUTH POLE PHOTOGRAPHS

A photograph has many advantages over a drawing. Discussed here are two photos taken in 1977, both at the U.S. research station in Antarctica, located right at the South Pole. The first is shown as Plate 4-1. Though it is often aesthetically unappealing to have an obstacle standing in the middle of a picture, the flag in the middle here reduces the intensity of the sun and greatly helps reduce lens flare and ghost images. Photographs taken without this compromise are likely to be flawed by a spurious ring or spots of light – artifacts of the bright sun in the field of the fisheye lens.

Plate 4-1 contains an interesting selection of effects that I think we understand fairly well. The simulation of that display should be a good test of how well our simulations do represent reality: The photograph was taken with a lens whose properties I have measured, and there is no question about all parts of the display having been recorded at the same time. The simulation of Figure 4-6 includes the

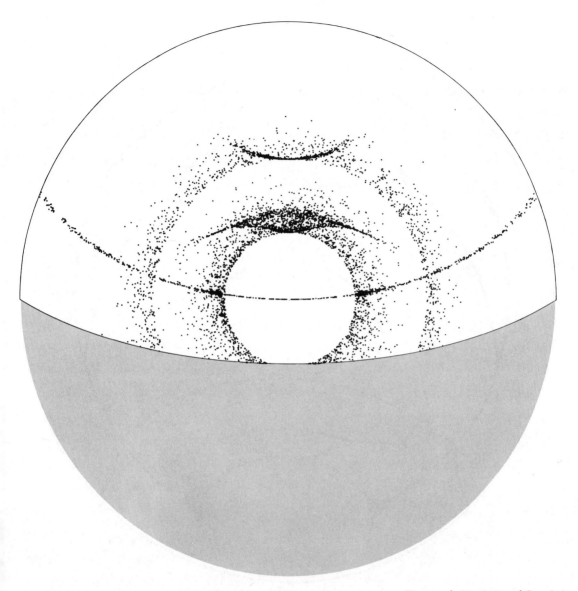

Figure 4-6. Simulation of Greenler's South Pole photograph of 1977 (Plate 4-1).

22- and 46-degree halos, 22-degree parhelia, parhelic circle, upper tangent arc, upper suncave Parry arc, and circumzenithal arc. I find the agreement most gratifying.

The photograph of Plate 4-2 was taken in February 1977 by Austin Hogan. A number of very interesting faint features are visible on the original and, as insurance against their disappearance in the reproduction, I include a drawing (Figure 4-7A) made from the photograph without reference to the simulation (Figure 4-7B).

The two halos, sun dogs, and parhelic circle are clear. The intense upper tangent arc to the 22-degree halo is accompanied by a strong upper suncave Parry arc. All these features are consistent with a sun elevation of 18 degrees, as measured from the photograph. If you

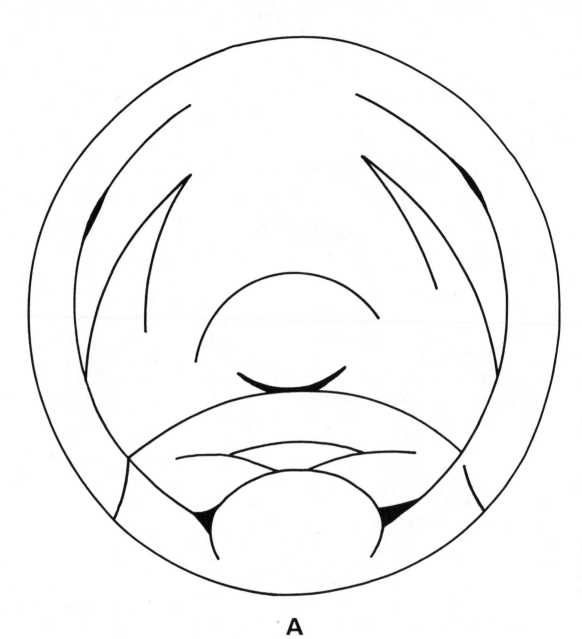

A

Figure 4-7. A: Drawing of effects
that can be seen in Hogan's South
Pole photograph of 1977 (Plate 4-2).
B: Simulation of Hogan's photo.

look closely at the 46-degree halo, you will see that it appears to
have a bit of a corner where it intersects the parhelic circle on the left
side of the sun. By my hypothesis, most of the intensity is actually a
supralateral and infralateral arc. This is consistent with the observation that the upper tangent arc, which comes from the same set of
ice crystals, is the most intense feature of the display. As you can see
from Figure 2-31, the supralateral arc coincides quite closely with
the upper part of the 46-degree halo for sun elevations near 20
degrees. The circumzenithal arc appears to be nearly tangent to this
supralateral arc, as is expected for this sun elevation. There is a faint

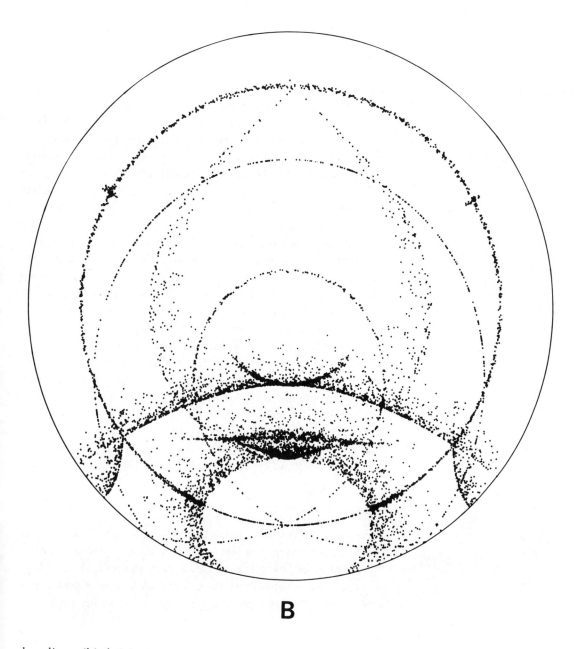

B

but discernible brightening on the parhelic circle at the 120-degree parhelic points, the one on the right being the brighter.

A column of ghost images marches across the picture from the sun to the anthelic point. Some are in the form of hexagons (a result not of the shape of ice crystals but rather of the six-sided iris aperture in the lens), and others near the anthelion are in the shape of flattened circles.

What remains are two *V*'s, to the left and right of the zenith, each with its apex pointing toward the anthelion, and an arc, concave toward the zenith, opposite the circumzenithal arc. That arc must certainly result from reflection off the side faces of the highly oriented pencil crystals responsible for the Parry arc. Their formation and appearance are illustrated in Figure 3-30, and the curve in the simulation just fits the observed portion of the curve in Plate 4-2.

Now let us consider the pair of V's. I do not know of an effect that produces a V shape in that part of the sky, but two superimposed arcs, both resulting from crystals that we know must be present, will produce the V. The anthelic arcs that result from Wegener's mechanism swing across the sky to form the inner arm of each of the V's. These arcs are formed by pencil crystals with long axes horizontal, though, actually, the Hastings anthelic arcs from the Parry-arc orientations of pencil crystals are so similar that we cannot distinguish them with certainty. The outer arm of the V is matched by the subhelic arc, which will be produced by either the Parry-arc or the circumscribed-halo distribution of the pencil-crystal orientations. It requires light passing through the horizontal pencil crystal as if it were a kaleidoscope – in one end and out the other with reflections off two side faces en route. In Plate 4-2 you can see an intensification where this arc crosses the halo about 70 degrees in azimuth from the sun. If this were reported as a "parhelion," its angular location would vary rapidly with sun elevation. There might be a set of Tricker's kaleidoscope anthelic arcs under all the ghosts at the anthelic point, though that is mostly speculation.

Finally, look at the simulation (Figure 4-7B). As you will certainly agree, it is an excellent match for the observation.

COMPLEX HALO PUZZLE

Plate 4-3 is a fisheye photograph that shows a wealth of effects (I fear that some of them may be lost in the reproduction process). Look at this photo to see how many you can identify and understand. Chapters 2, 3, and 4 should have contributed both to your observation and to your understanding; check the Appendix for my listing.

5

Scattering: light in the sky and color in the clouds

Of course the sky is dark at night; perhaps the thing that should surprise us is that it is light during the day. On a visit to the moon you would experience day and night (though the full earth would provide light at night), but during the daytime on the moon, the sky is dark. Where, then, do we get our blue sky?

LIGHT SCATTERING

If we look in a skyward direction, away from the sun, and see blue sky, it means that light is coming from that direction in the sky to our eyes. Light from the sun has changed its direction somewhere in the sky; otherwise, all we would see is the dark of outer space or the light from some distant star. When a ray of light passes through our atmosphere, some of its energy is deflected in all directions by a process called scattering. The moon's sky is black because it has no atmosphere to scatter sunlight.

BLUE SKY

Sunlight scattering by the air can explain why the sky is illuminated, but to understand why the clear sky is blue and why clouds are white, consider the wave nature of light. All kinds of waves are scattered by obstacles in their path. A boulder will scatter water waves, sending wavelets off into new directions; a slender reed will have little effect on a water wave, which surges by as if the reed were not there. A wave will be scattered significantly only if the obstacle is "big enough," and the scale by which we can determine what is "big enough" is the wave itself. An obstacle that is very small compared to a wavelength will not be an effective scatterer. The same obstacle will scatter more effectively for waves with shorter wavelengths, because the obstacle will be relatively larger. This is the situation with light scattering in the atmosphere; the scatterers are smaller than the wavelengths of visible light.

Air scatters light because it is not a continuous fluid but is composed of discrete molecules. If we consider the molecule itself as the scatterer, then it is about a thousand times smaller than the wavelength of light. Another model considers that the molecules do not stay uniformly spaced but, as they bounce around colliding with each other, are sometimes temporarily bunched up in certain small volumes. These statistical fluctuations in population density constantly occur and provide the lumpy structure in the air that causes light scattering. From this statistical model I would estimate that the lump size is more like the typical distance between air molecules — which is still a hundred times less than the wavelength of light. Without going deeper into the problem, we can conclude that the scatterers are smaller than a wavelength of visible light. A consequence is that they will scatter the shorter (blue) wavelengths more effectively than the longer (red) wavelengths. There is our explanation: The light we see in the sky is scattered light, and the atmosphere scatters the blue light more effectively than the red; so the clear sky is blue.

Most of the colors we see in nature are pigment colors, produced by materials that selectively absorb certain wavelengths of light. The blue of the sky is different from this: There is not really any blue material in the sky, just small particles. This effect is not unique to the sky. The solid particles in milk are smaller than a visible wavelength; hence the bluish color when you rinse out the milk glass. It has also been claimed that the blues in the feather of a bluebird and bluejay are not pigment colors but result from a very fine scattering structure in the feathers.

WHITE CLOUDS

We also see clouds by scattered light, and clouds, as everyone knows, are usually white. Clouds are made up of small water droplets or of ice crystals. But the "small" size of cloud droplets is quite a different "small" from that of molecules. In fact, the small cloud droplets are large compared with wavelengths of visible light — typically on the order of fifty times larger. These relatively large drops scatter all wavelengths of light equally well, giving them the appearance of being white. What we see as white in clouds is actually a collection of small, transparent, colorless drops of water. So, to complete the paradox, not only is there nothing blue in the blue sky, but there is nothing white in the white clouds.

A common example of this same effect can be seen in the smoke of a cigarette or cigar. The smoke rising from the burning end has a bluish appearance and the smoke from the other end is distinctly whiter. It is a case not of blue particles and white particles, but rather of smaller particles and larger particles.

The blue-sky blue, as just noted, is not a pure spectral color but contains all of the spectral colors in decreasing proportion from blue to red. As the blue light is selectively removed by scattering from a beam of sunlight passing through the atmosphere, the color of the unscattered beam must also change. The light passing through the atmosphere first takes on a yellowish tint, which changes to orangish and to reddish as scattering removes more and more of the short wavelengths of light. This is the progression you see in the color of the sun over the course of an afternoon. As the sun moves lower in the sky, the length of the path through the atmosphere increases for light rays that get to an observer's eyes. As the sun sets, the long slant path of the rays results in so much light scattering that the setting sun looks red. Neaby clouds are illuminated with reddish light, and as the nearby atmosphere is illuminated by redder and redder light, even the scattered light becomes reddish.

The blue sky and the red sunset are, therefore, actually two complementary aspects of the same phenomenon. I recently noticed another demonstration of the complementarity of these effects. I visited the cabin of some friends on a cold, sunshiny winter day. When we got a hot wood fire going in the small stove, I saw that the smoke coming out of the chimney looked blue in the bright sunshine, but that its shadow on the snow was quite reddish. You can also see the effect by adding ten or fifteen drops of milk to a glass of water and seeing the bluish color; and then looking through the glass at a light bulb and seeing that the bulb, viewed through the milky water, is considerably reddened.

The redness of a sunset can be increased by the addition of small scattering particles to the atmosphere. Frequently we refer to this process as air pollution, but significant natural variations occur in the number of small particles in the atmosphere. Large forest fires in the western part of the United States put enough fine smoke particles into the atmosphere to intensify the sunsets over the rest of the country as the prevailing winds carry the particles eastward. Major volcanic explosions project fine particles so high into the atmosphere that they circle the earth many times with a lifetime in the atmosphere of a few years before settling. The effect of a large volcanic eruption can be seen in intensified sunset colors all over the world.

THE COLOR OF THE NIGHT SKY

Have you ever wondered what color the sky is at night? Plate 5-1 shows a view, including the sky, taken in midafternoon. Plate 5-2 shows the same scene photographed about 10:30 that night. I took both pictures with the same film but gave the second an exposure ten million times greater than the first. The main source of light in the second picture was the full moon. Because moonlight has almost the same color as direct sunlight, you would expect the colors of the photographs, including the blue sky, to be about the same. There is a slight shift in the color balance of the picture resulting from a peculiarity of the film when it is used for long exposure times. This effect varies between different kinds of color film and probably accounts for most of the color difference between the two pictures. But the answer to the question is clearly that the sky is also blue at night (if it is scattering white light).

BLACK CLOUDS

We think of clouds as white; yet their color can vary widely. Let us consider a few of the possibilities. Plate 5-3 shows a combination of what appear to be white clouds and black clouds. However, the dark cloud is really far from black: It is gray by comparison with the lighter cloud behind it. Another, even brighter, cloud in the picture would make the present "white" cloud appear gray. Figure 5-1 schematically shows identical gray clouds on light and dark backgrounds. This gray figure appears brighter on the darker background. What we identify as white is just the brightest gray in sight; there is no standard that separates whiteness from grayness.

Figure 5-1. The background affects the apparent brightness of the gray figure.

There are two different reasons why we might look at two clouds and call one white and the other black (or gray). One of the clouds may be less bright than the other because it is in a shadow of another cloud; this is probably the case in Plate 5-3. This is the same effect that can make a dense cloud appear white on the top but dark underneath, where it is shadowed by the upper part of the cloud. Or we may see two similar clouds against backgrounds of markedly different brightness, as in Figure 5-1, and judge one to be white and the other gray. Plate 5-4 is a photograph of such a situation, taken as I watched a line of clouds moving across the sky. The same cloud that, in the foreground against a bright sky, looks dark appears to be

white as it moves into the left background, against a dark part of the
sky.

CLOUDS OF DIFFERENT COLORS

The cloud of Plate 5-5 is yellow near its base for reasons that you now understand. It is illuminated by late afternoon sunlight, which is yellowish because of the selective scattering of blue light. Plate 5-6 shows the same effect for a lower sun where the unscattered light is redder. In many cases the detailed color of clouds is a complicated combination of the color of the light that illuminates them (determined by its scattering history) and the droplet size (hence, the scattering properties of the clouds themselves). Our impressions of the color are further influenced by the color of the background against which they are viewed. Plate 5-7 shows a blue cloud. (Although I cannot document a good green cloud, I believe that they also do occur).

AIR LIGHT

Plate 5-8 is a photograph from the Great Smoky Mountains. It is reasonable to assume that the trees and bushes on all the mountains, near and far, are the same color; and yet, as you look from the foreground to the next mountainside and on to those mountains farther in the distance, you see them increasingly submerged in a blue haze. The blue light that reaches your eye as you view such a scene is sunlight, which is scattered toward you by the air between you and the mountain. The greater the distance, the more light-scattering air there is between you and the distant mountain. This scattered light is blue, of course, and produces the familiar blue color of distant mountains. (The fact that the haze gets whiter for the most distant peaks is a subtle effect, which I will discuss later in the chapter.) We unconsciously use this air light to judge distances. When we are in a region with exceptionally clear air, either at a high elevation, where the atmosphere is thinner, or at a place where the air is cleaner, we tend to underestimate distances.

CREPUSCULAR RAYS

Another way to see light scattering in the atmosphere is to isolate a beam of light, which is visible from the side only because some light is scattered out of the beam. Figure 5-2, showing the sunlight streaming through the cupola windows of St. Peter's Cathedral in Rome, illustrates this familiar effect. Outdoors, the function of the window in isolating beams of sunlight is sometimes provided by

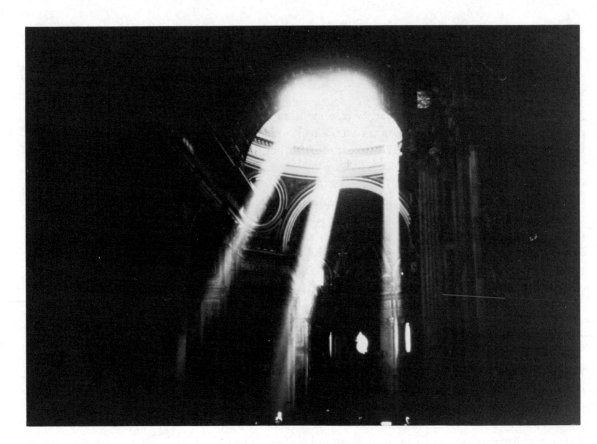

Figure 5-2. Sunlight coming through cupola windows of St. Peter's Cathedral. (Photographed by Howard A. Bridgman)

openings in a cloud layer (Plate 5-9). These beams, passing through the clouds, are called crepuscular (literally, twilight) rays. A common folk description of the effect illustrated in Plate 5-9 is the expression "The sun is drawing water." I suppose this comes from an early attempt to explain how water returns to the sky after a rain, so that it can rain again.

Another example of the same effect is shown in Plate 5-10, where the rays appear above the sun. Here it is reasonable to wonder why the rays in this late afternoon picture radiate as they do. If the sun were a big light bulb located just behind the trees, then it would seem right to see rays radiating upward from it in all directions. But a scale drawing of the earth at its great distance from the sun would show that the sun's rays are nearly parallel as they pass through the earth's atmosphere, though they do not look parallel in Plate 5-10. A key to this puzzle is given by the photograph (Plate 5-11) of railroad tracks trailing off into the sunset. Looking at this picture, you immediately assume that the tracks are parallel, even though they appear to converge to a point on the horizon. Other parallel tracks on either side of these would converge to the same point, and if there were telegraph lines overhead, they would enter the picture from the top and converge to the same vanishing point on the horizon. That is the familiar perspective of parallel lines. The same perspective, applied to parallel beams of sunlight passing through our atmosphere, gives the sunburst effect of the crepuscular rays. Those rays radiating above the sun are rays that pass overhead, whereas those below the sun will hit the earth in front of the observer.

Although I have described crepuscular rays as isolated beams of light passing through the atmosphere, dark rays can be formed in the same way by the projection of shadows. A tree or cloud may leave a shadow streak across the sky. What is needed to see the scattering is a boundary between air that is illuminated and air that is not. Plate 5-12 shows a spectacular view of this effect. Rays from the setting sun shine through the trees, with the scattering enhanced, in this case, by ground fog.

ANTICREPUSCULAR RAYS

As you follow a crepuscular ray away from the sun, overhead it becomes fainter: There is more scattered light reaching your eye when you look along a thin beam of light than when you look across it at right angles. Sometimes, however, the rays can be seen as they pass overhead and beyond. If you have never actually checked to see what form they take in the process, you might guess — incorrectly — that they continue to fan out, spreading ever wider across the sky. Let us go back to the parallel railroad tracks, which appear to diverge as they approach you from the distant horizon. Consider how the tracks appear when you turn around: They converge to the opposite horizon. Similarly, these parallel rays from the sun converge to the antisolar point. Remember that if the sun is above the horizon, the antisolar point will be below the horizon on the opposite side of the sky, and so the crepuscular rays should converge to a point below the horizon. Such rays, called anticrepuscular rays, are well illustrated in Plate 5-13.

THE TWILIGHT WEDGE

On clear, smog-free days, shortly after the sun has set in the west, a band of the sky next to the eastern horizon lies in the shadow of the earth. The boundary between the illuminated sky and the shadowed sky appears fairly sharp when the sun has just disappeared below the horizon, but as the sun sets, the boundary rises and becomes more diffuse. The shadow boundary is highest on the side of the sky opposite the sun and tapers to the horizon on the northern and southern sides of the sky; its shape suggests the name by which this effect is sometimes known: the twilight wedge. The drawings of Figure 5-3 help explain why the twilight-wedge boundary becomes diffuse

131

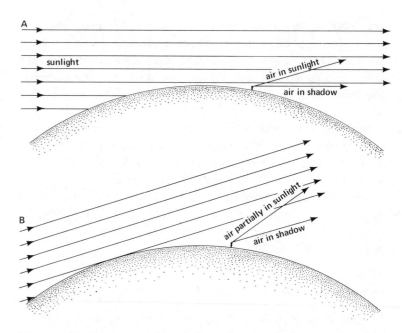

Figure 5-3. A: The twilight-wedge boundary is relatively sharp immediately after the sun has set. B: As the boundary rises it becomes more diffuse.

as the sun sets. In Figure 5-3A, the observer is looking right along the boundary; one line of sight sees only air in shadow, and a slightly higher line of sight sees mostly air illuminated by sunlight. In Figure 5-3B, however, as the boundary passes high overhead, the proportion of the air path in sunlight increases slowly as the line of sight is raised.

The photograph of Plates 5-14 and 5-15, taken a few minutes apart, show the rising twilight wedge. Notice that just above the shadow line there is a reddish band of light, corresponding to illumination of the distant air by sunlight robbed of its blue component by atmospheric scattering. From Figure 5-3B you can see that, were you to observe the twilight wedge from an airplane, you should be able to see a sharp boundary above the horizon. For example, in a plane flying at an altitude of 10,000 meters, you would be at the light – dark boundary, and hence would see the sharpest earth – shadow line, when the line was about 3 degrees above the eastern horizon.

THE MOON'S SHADOW ON THE EARTH'S ATMOSPHERE

The moon casts its shadow on the earth's atmosphere when the moon is between the earth and the sun, producing a solar eclipse. Figure 5-4 shows such an event, shortly after the totally eclipsed sun has risen. The moon is the small dark disk centered on, and entirely covering, the sun. The surrounding brightness is the sun's outer atmosphere; although it appears bright in this long-exposed photo, it is too faint to be observed except during a total eclipse. The shadow of

the moon does not cover the entire earth but covers the spot where the observer was located. As you look in a direction near to the eclipsed sun, your line of sight traverses only atmosphere lying in the moon's shadow; however, if you look far enough away from the eclipsed sun, you can see atmosphere that is illuminated by the sunlight and appears bright by scattered light.

Figure 5-4. Photograph of the sky during a total eclipse of the sun ca. 1950. (Contributed by John Strong; photographer unknown)

THE ANGULAR INTENSITY VARIATION OF SCATTERED LIGHT

The angular variation in the intensity of scattered light depends on the size of the scattering particles. The simplest case is the blue-sky scattering in which the scatterers are very small compared with the wavelength of the scattered light. This is called Rayleigh scattering, after Lord Rayleigh, whose wide-ranging curiosity led him to contribute understanding to this subject and to dozens of others in the late 1800s. Rayleigh was able to calculate the relative intensity of light scattered by clean air in all directions. In Figure 5-5A, arrows represent light scattered in different directions, with the length of each arrow proportional to the intensity of the light scattered in that direction. This information can be represented more conveniently by

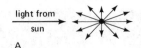

A

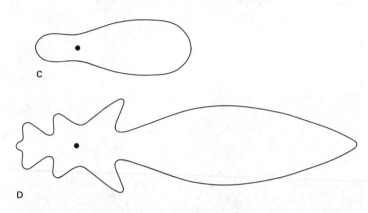

B

C

D

Figure 5-5. A: Schematic representation of the angular variation in the scattered intensity from very small particles; for example, particles smaller than one-fiftieth the wavelength of light. B: Another way to represent the light-scattering pattern in A. C: Scattering pattern for particles approximately one-quarter the wavelength of light. D: Scattering pattern for particles larger than the wavelength of light.

drawing the curve connecting the heads of the arrows, shown in Figure 5-5B. On this plot, the intensity of scattered light in any direction is measured by the distance from the center to the curve in that direction; scattering in the forward and backward directions has the same intensity, which is greater than the scattering at right angles only by a factor of two. This does not agree with our observation that the sky near the sun direction is usually much brighter than the rest of the sky. The conclusion to draw is that we usually do not see only Rayleigh scattering in the sky, but see also scattering from larger particles.

Theories developed after Rayleigh's enable us to calculate the angular scattering distribution for particles of different sizes. Figure 5-5C and 5-5D show the scattering plots for larger particles, and we can see that as the particle size increases, there is much more light intensity in the forward direction — light observed by looking near the direction of the sun. These diagrams give the basis for a simple rule by which to judge when the atmosphere is really clear, that is, when there is mainly Rayleigh scattering from the air, without larger-particle scattering. Close one eye and block out the sun with your upraised thumb, held at arm's length. In a very clear atmosphere, the sky right next to the thumb-obscured sun appears no brighter than the sky some distance away. Plate 5-16 is a photograph of a sky so clear that it passed my thumb test. I took it at the U.S. Research Station at the South Pole in Antarctica. My head, rather than my thumb, blocks out the sun, and you can see that there is no obvious increase in sky intensity close to the obscured sun. Contrast this fisheye photograph with another (Plate 5-17), taken near Milwaukee on a day with a slight haze. The increased forward scattering can be seen clearly as an enhanced sky brightness near the sun.

Several other processes affect the distribution of the light intensity over the blue sky. Note that in Plates 5-16 and 5-17 the sky is brighter and whiter toward the horizon than toward the zenith. The most important effect is the variation in the thickness of the atmosphere for different elevations of your line of sight. You look through a much longer air path in the horizon direction than in the zenith direction. If you find it surprising that increasing the scattering path makes the scattered light whiter, think of it this way: Suppose light is shining from above onto a long horizontal tunnel of air whose length we can arbitrarily vary. As you look down this tunnel, all wavelengths of light are scattered in your direction, but more blue than red light is scattered. However, the blue light that is scattered in your direction from the far end of the tunnel gets selectively attenuated by additional scattering on its path toward your eye, whereas the red light coming your way is attenuated much less. As the tunnel of air is made longer and longer, those two opposing effects cancel, leaving the scattered light white again, in the limit of a very long air path. This is the effect that, even on a clear day, makes the sky whiter near the horizon than at the zenith. It is also the effect, evident in Plate 5-8, that makes the most distant mountains look whiter than those at moderate distances.

The skyline of Paris is shown spread out in the panorama of Plate 5-18. A thin smoggy layer lay over Paris on this December day, and the observation level on the Eiffel Tower was located at just the right height to be in that layer. When I went down to the next lower level, I was out of the layer and could not see the effects I describe here. In the panel on the right, the shadow of the tower marks the direction opposite the sun. Viewed in that direction, the smoggy layer is much darker than the clearer sky above it. In terms of the scattering curves of Figure 5-5, it means that the backscattered light intensity is less for the larger smog particles than for the blue sky. As we look from right to left, moving closer to the direction of the sun, the blue sky gradually brightens; the smog layer stays dark until, in the second panel from the left, its intensity increases very rapidly, becoming brighter than the blue sky above. We can see in a dramatic way the differences in angular scattering of the larger smog particles, perhaps represented by Figure 5-5D, and of the smaller, blue-sky particles, perhaps represented by a combination of Figure 5-5B and C.

Most of the effects discussed in this chapter are relatively common and can be seen by anyone who looks. The exceptions to this generalization are perhaps:

1. Rayleigh scattering from a really clean atmosphere
2. The haze layer seen from the Eiffel Tower — although the effect can frequently be seen from an airplane
3. The scattering effect seen during a total solar eclipse.

THE ICE BLINK AND WATER SKY

The discussion in this section concerns an effect that most of us have never seen and that has been reported mostly in polar regions. I suspect, nonetheless, that related effects are visible to the aware observer in other parts of the world.

When the sun shines down through a thin layer of clouds, some of the light is scattered to the observer's eye, and by it, he or she sees brightness in the clouds. The downward-scattered light strikes the earth and is again scattered, and this earth-scattered light again impinges on the cloud layer. The nature of this ground-scattered light can affect the appearance of the clouds, particularly when the ground is covered with white snow or ice so that little of the light is absorbed, most of it being rescattered. The term that describes the fraction of the incident light that is rescattered back from the earth's surface is the *albedo*. A perfectly black absorbing surface would have an albedo of zero; on the other end of the scale, a perfectly white, diffusely reflecting surface would have an albedo of one. For a surface with a high albedo, the brightness of the cloud cover may be increased not only by light scattered once from the ground but by light that is successively scattered several times between ground and cloud.

Let me digress here to point out a closely related effect. In the situation where light is repeatedly scattered between cloud and snow, the flux of light traveling upward or downward at any angle is the same as that at any other angle. In such a uniform white box, no shadows exist and even the horizon line disappears. This is the whiteout described by Arctic and Antarctic travelers in which all visual cues of orientation and ground structure are gone. Lacking such visual cues, some people have difficulty even in maintaining their balance; travelers stumble, unseeing, into ice blocks or snow drifts or step off invisible ledges in the featureless snowscape.

What I want to concentrate on is not the effect of this multiple reflection on the appearance of the ground, but rather the effect on the appearance of the cloud layer. If two adjacent ground areas have drastically different albedos, this difference will be apparent in the brightness of the sky above them. Thus a map of bright and dark ground features may be seen projected onto the cloud cover. In fact, this very explicit cloud map is reported to be used by Eskimos in traveling either by boat or on foot across sea ice. When traveling by boat, the Eskimo interprets a white *ice blink* in the sky as evidence of

sea ice that can be avoided before it is sighted directly. When traveling across the ice, he interprets the dark *water sky* as evidence of
leads – cracks in the sea ice – to be avoided.

Catchpole and Moodie[1] have described the effects of such multiple reflections in Arctic regions, and they cite two references that
describe the phenomenon. Writes Stefansson[2] in *The Friendly Arctic:*

> Now we had a good deal of cloudy weather and found the
> "water sky" exceedingly useful. When uniformly clouded over
> the sky reflects everything beneath it in the manner of a mir-
> ror. If there is below a white patch of ice, then the sky over it
> looks white, while a black strip of water is represented by a
> black line in the sky. It is hard on the eyes to travel in cloudy
> weather and hard on the dogs for picking trail, yet the water
> sky absent in clear weather more than makes up for these
> disadvantages. Leads were all about us but the corners of
> various cakes were touching, and by keeping our eyes on the
> cloud map above we were able to travel sometimes a day at a
> time without even seeing water.

The second reference is to Scoresby's[3] geography of arctic regions:

> On approaching a pack, field, or other compact aggregation of
> ice, the phenomenon of the ice-blink is seen whenever the ho-
> rizon is tolerably free from clouds, and in some cases even
> under a thick sky. The ice-blink consists in a stratum of a lucid
> whiteness, which appears over ice in that part of the atmo-
> sphere adjoining the horizon . . . when the ice-blink occurs
> under the most favourable circumstances, it affords to the eye a
> beautiful and perfect map of the ice, twenty or thirty miles
> beyond the limit of direct vision, but less distant in proportion
> as the atmosphere is more dense and obscure. The ice-blink
> not only shows the figure of the ice, but enables the experi-
> enced observer to judge whether the ice thus pictured be field
> or packed ice: if the latter, whether it be compact or open, bay
> or heavy ice. Field-ice affords the most lucid blink, accom-
> panied with a tinge of yellow; that of packs is more purely
> white; and of bay-ice, greyish. The land, on account of its
> snowy covering, likewise occasions a blink, which is more
> yellow than that produced by the ice of fields.

OTHER EARTH–SKY SCATTERING EFFECTS

We are left with the question whether there are some circumstances
outside the polar regions where we can derive from the sky optical
information about the ground surface. I suggest here a short list of
possibilities and leave it to the reader to extend that list.

Many years ago I spent a couple of days on a long narrow barrier island off the east coast of the United States. The island was about a kilometer wide, the half facing the Atlantic Ocean being a white sand beach, the other half a salt water marsh. In the slightly hazy air just above the beach, I noticed an intense sky brightness as I looked either up or down the length of the beach. The intensity dropped off abruptly either toward the ocean or toward the marsh. At the time I had never heard of the ice blink and had only some vague thoughts of a concentration of dust particles over the hot beach. Now I would guess that I was seeing what I might call a "sand blink."

I have heard it suggested that a cloud can have a greenish cast as a result of light scattered upward from green vegetation. If so, the effect might be identified somewhere where there is a large block of a bright green field crop adjacent to plowed land, or to nonirrigated land, or to a mature crop that is no longer green.

Sometimes a late spring or early fall snow falls on ground that is not frozen and the snow is melted by heat from the ground. (In the warmer parts of the country, where snow is rare, this may be the usual situation.) The bare fields may lose snow cover before fields insulated with a layer of vegetation, and the result might be large ground areas of drastically different albedos. Such variation might be visible in an appropriately overcast sky.

Another possibility is evidence in the sky of a bare blacktop road or of a river when the rest of the countryside is covered with snow.

LIGHT-SCATTERING PUZZLE

The puzzles for this chapter are shown in Plates 5-19 and 5-20. These rainbows show some light-scattering effects that should be understandable from the discussion in this chapter. Why not try to identify and explain the effects before you check my comments in the Appendix?

6

Diffraction: the corona, the glory, and the specter of the Brocken

If you look at the full moon on a misty night, you see it surrounded by soft rings of light. On a night with broken clouds, drifting across the sky, you see them transformed as they move into the sphere of influence of the moon, momentarily becoming part of a beautiful circle of light before fading and moving on. Shift your attention from the clouds to the moon's circle of light and see it expand and shrink, strengthen and fade, reflecting the individuality of the passing clouds. This is the corona (literally, crown), produced by small water drops or ice crystals in the earth's atmosphere.

THE CORONA

Do not confuse the corona with the much larger 22-degree halo. The diameter of the circle of light or of the rings that make up the corona display is a few moon diameters — much smaller than the halo. The most common form of the corona is a disk of light, bluish near the moon, becoming whiter farther out, and terminating in a reddish brown outer edge (Plate 6-1). It is called the aureole and can be observed in a variety of sizes and intensities. The display becomes more impressive when the aureole is surrounded by a colored ring, or sometimes a few such rings. Plates 6-2 and 6-3 are photographs of such corona displays appearing about the sun rather than the moon, but the physics of the display is the same for either source of light. Many people see the moon corona more often because the sun's brightness makes it difficult to view effects located near the sun direction. Here is where a water puddle can help. A water surface reflects only about 2 percent of the light falling on it. When you look at the sun corona reflected in a puddle, you benefit from a fifty-fold reduction in intensity, which frequently allows comfortable viewing. Alternatively, you might use the reflection from a house or car window, or view the scene through dark sunglasses.

Another, unrelated effect may be confused in name with the one I am discussing. The outer part of the sun's atmosphere is called the corona and is visible when the moon blocks out the sun's disk during a total solar eclipse; Figure 5-4 shows this part of the sun's atmosphere. The effect that I am discussing here, however, arises from clouds in the earth's atmosphere and has nothing to do with the atmosphere of the sun.

DIFFRACTION OF LIGHT PRODUCING THE CORONA

The corona rings (and most of the effects I discuss in this chapter) are diffraction effects that result from the wave nature of light. In explaining the nature of the white rainbow I mentioned that light from a small source will pass through a round hole and produce a sharply defined, circular spot of light on a wall beyond. The size of the spot on the wall is proportional to the size of the hole, until the hole is made very small. For a very small hole, the light beam broadens as it passes through the hole, giving a larger, more diffuse spot than would result from light rays traveling in exactly straight lines through the opening. Actually, if the experiment is done carefully, using light of only one color (monochromatic light), the diffuse spot is surrounded by a series of concentric circles of light that decrease in intensity from one ring to the next larger. The diameters of the rings depend not only on the size of the hole but also on the wavelength of the light. Thus the pattern of rings is larger for red light than for blue.

According to a principle of optics, we can treat every small area inside the circular hole as a source of wavelets traveling off in all directions. If you add up the results of the wavelets from all parts of the hole you find that in some directions they cancel one another (destructively interfere), and in other directions they reinforce one another (constructively interfere). On the wall, these directions show up as dark rings and light rings surrounding the diffuse central circle of light.

A very similar diffraction phenomenon occurs when light from a small source is allowed to fall on a wall after we replace the cardboard containing the small hole with the small circular disk cut out of the cardboard. The small, opaque circular disk casts a diffuse circular shadow surrounded by diffraction rings on the otherwise illuminated wall. One of the most surprising aspects of this phenomenon is that the light intensity on the wall at the bright rings surrounding the shadow is brighter than it would be if the opaque disk were removed. By blocking out some of the light we actually make certain areas on the wall brighter.

The corona is the diffraction pattern of moonlight (or sunlight) caused by water droplets or ice crystals in the atmosphere. In explaining the rainbow, halos, and sun pillars, I started out by considering light reflected or refracted by one particle. The pattern produced by light falling on a screen from one particle, given many different orientations, has turned out to be just what an observer should see by looking at a sky filled with many particles. This is also the case for the corona. For instance, if red light is deviated by 2 degrees into the first bright diffraction ring, whenever the observer looks in a direction 2 degrees away from the sun, he or she should see red light, specifically, a red diffraction ring of 2-degree radius around the sun. Figure 2-6 illustrates the equality between the deviation angle and the observing angle.

COLORS OF THE CORONA

Suppose we have the sky filled with uniform-sized drops and we look only at the red-light diffraction pattern. In the sky we should see a diffuse circle of red light, centered on the moon, surrounded by red rings. If we look at the blue light we should see the same pattern in blue, shrunk to about two-thirds of the red-pattern size. If we look at both the red- and the blue-light patterns superimposed, we will see the central circle bluish toward the center (where the blue light is concentrated) and red toward its outer edge. If we add all of the spectral colors in between the red and blue, we expect to see a nearly white light between the bluish inner part and the reddish outer part of the central disk of light. That is a rather good description of the aureole of a corona display. The rings surrounding the aureole also have blue inner edges and red outer edges. For higher-order rings the spread of colors may get so large that the red of one ring overlaps the blue of the next larger ring, giving a mixture of colors not found in a pure spectrum.

Remember that the size of the rings depends not only on the wavelength of the light but also on the drop size, with smaller drops producing larger rings. If there is a mixture of widely different drops sizes in a cloud, the corona we see is the superposition of a number of corona patterns of widely varying sizes. The result is that the intensity and color variation in the rings blur into a uniform illumination and only the aureole remains. This is probably the situation for the photograph in Plate 6-1, whereas the drops were of more uniform size in Plates 6-2 and 6-3.

CLOUD IRIDESCENCE

Sometimes the size of particles may change from one part of a cloud to another. In Plate 6-4 you see a magnified portion of the corona, showing how variation in droplet size affects its shape. Small-scale variation give a roughness associated with the streak clouds, and larger-scale variation distorts the general circular shape. As a cloud evaporates, the droplets along the edge disappear more quickly than those inside the cloud, so that in an evaporating cloud there are smaller droplets toward the edge. We could use the angular radius of the first red ring (first-order red light) from the sun to calculate the size of the water droplets in that particular part of the cloud. If the sun is shining through a layer of clouds in which the drop size changes only slightly from one part of the sky to another, then the diffraction rings will be slightly distorted circles. If, however, the drop size changes drastically over a short angular distance in the sky, as in a small, evaporating cloud located in a direction a few degrees from the sun, the picture becomes quite different. The whole cloud is approximately the same angular distance from the sun, and the line of first-order red light traces out a contour on the cloud in which all of the drops are of about the same size. The first-order blue light appears on another cloud contour where the drops are smaller.

The diffraction colors you see in clouds are called iridescence. In many iridescent clouds the colors are pastel pinks and blues or greens. Plate 6-5 shows such a cloud, with the color lines appearing nearly as contour lines around it. The pastel colors may indicate that we are seeing a higher order of diffraction, in which colors of one order are overlapping those of another. Plate 6-6 shows the iridescence near the sun where the colors are more saturated (more like spectral colors). In this picture, taken with a telephoto lens looking just to the left of the sun, you can see that the color pattern is determined both by the angular distance from the sun and by the variations of drop size.

The sky around the iridescent cloud of Plate 6-6 appears almost black in the photograph. Actually, it was a bright, winter-day blue, and that difference suggests why you may not have seen such striking displays before. The iridescent cloud near the sun is very bright, so bright that when it is correctly exposed in a photograph, the blue sky is drastically underexposed. It is so bright as to dazzle your unaided eyes. I first became aware of iridescent clouds many years ago after several canoe trips during which I spent time looking at the sky reflected in still water. Cloud iridescence is really very common, occurring most days when broken clouds move across the sky.

BISHOP'S RING

Thus far we have considered diffraction rings around the sun or moon arising from water drops or ice crystals in the atmosphere. The presence of any kind of particles in the atmosphere will give observable diffraction effects if they are small enough, sufficiently uniform in size, and present in high enough concentration. Major volcanic eruptions sometimes inject tremendous quantities of solid particles into the atmosphere. The larger particles settle out quickly, leaving fine volcanic dust circulating in the earth's atmosphere over a period of years. The most spectacular volcanic event about which we have extensive recorded observations was the eruption on the island of Krakatoa near Java. Major explosions occurred in August 1883, and over the following years the worldwide effects were so dramatic that the Royal Society of London (the most prestigious scientific group of the time) appointed a committee to gather observations on the effects from around the world. Their report, *The Eruption of Krakatoa and Subsequent Phenomena,* is a five-hundred-page book; one of the sections of this report[1] deals with a large corona around the sun and the moon associated with the spread of the volcanic dust. After the explosion on August 27, the first published observation of the corona came from the *Japan Gazette,* which described the appearance of a "faint halo" around the sun on August 30. The first detailed observation of it was made by the Reverend Sereno Bishop at Honolulu on September 5. His description of the corona was accompanied by his observation of increased light scattering in the twilight sky, a result of scattering particles so high in the atmosphere that they were still illuminated some time after sunset for earthbound observers. These observations established the phenomenon as identical to the one that appeared in Europe and America more than two months later. As a result of these first careful observations, this corona phenomenon has since been called Bishop's ring.

Although subjective descriptions of the colors of the ring differed widely, observers generally agreed that the inside of the ring was whitish or bluish white and that it shaded to a reddish or brownish or purplish outer edge. The sequence of colors, with the red on the outside, identifies it as a diffraction corona rather than a refraction halo (e.g., the 22-degree halo), for which the red is on the inside of the ring. The angular radius of the red ring, as averaged from estimates and measurements of several observers, was about 28 degrees. This is very large for a diffraction ring and indicates the very small size of the dust particles (about 0.002 millimeter). Although I have never seen Bishop's ring, I wonder if it occurs as a result of some of the man-made pollution in our skies.

THE SPECTER OF THE BROCKEN

Consider the following scenario, taking place at an earlier period, when most travel is by foot or on horseback. You have arrived at a mountain summit after a long climb, there being no other way to experience the thrill of gazing at a grand expanse of the earth's surface. You have labored under an overcast sky that became brighter and brighter as you climbed, until you broke through the clouds into the bright sunlit world of the summit. Now you stand on the mountain top, looking out over a world largely hidden from your view by the clouds. As you survey your surroundings, you see a giant shadow figure standing in the cloud mist. The apparition moves as you move and is clearly some magnified projection of yourself in this very special place. Around its head – your head – you see a series of brilliantly colored halos.

How you respond to this spectacle depends on your experiences and personality. Some have interpreted it in religious terms; others have searched for a physical explanation. From its frequent appearance on the Brocken, the highest peak in the Harz Mountains of central Germany, this display has been called the specter of the Brocken, and the colored rings have been referred to as the Brocken bow. Of course, it is not unique to that mountain: The photograph of Plate 6-7 was taken in the Jura Mountains between France and Switzerland.

There are two separate parts to the specter of the Brocken: the shadow and the colored rings. I do not fully understand why the shadow appears greatly enlarged to some observers, but I will speculate on the effect nevertheless. When you look at an object, you unconsciously form an opinion of its real size by combining two observations: its angular diameter and its distance from you. Thus, if you see two objects, each of which spans 1 degree in your field of view, you may perceive them as being of drastically different sizes, depending on how far away they appear to be. If one of the objects appears to be fifty feet away, you will judge it to be the size of a football; if the other appears to be a kilometer distant, you will perceive it to be the size of a house. You would make the same kind of judgment about the size of a shadow cast on a solid surface, but in the specter, the shadow is not localized at a particular surface but rather is projected through the fog. The size impression you get then depends totally on your estimate of the distance between you and this unfamiliar type of shadow. Perhaps those people who see their shadows as vastly enlarged have unconsciously judged them to be located at a great distance.

The specter shown in Plate 6-7 is clearly the shadow of the photographer, and yet it has a strange triangular appearance. The explanation lies in the perspective effect illustrated by the converging railroad tracks of Plate 5-11. Imagine the sun's rays that just graze your body in passing – those rays that outline your figure – projected as parallel lines into the fog before you. They all converge to the antisolar point, which is marked by the shadow of your head. The fact that many of the shadow rays in Plate 6-7 appear to converge to the head indicates that we are seeing the shadow projected a long way through the fog.

Now let us consider the colored rings. There are two directions in the sky where we can observe the effects of diffraction by small drops. Looking toward the sun, through a thin cloud, we see it surrounded by diffraction rings – the corona. We can also see diffraction effects in the opposite direction: Looking away from the sun, where the sunlight falls on a cloud, we can see the antisolar point surrounded by diffraction rings. These rings form the bow about the head of the specter of the Brocken.

If two people stand together on the summit of a mountain, both can see two shadows, though each sees the shadow of the other as relatively indistinct. But each sees only one set of rings, and sees it around the shadow of his or her own head.

THE GLORY

Before the days of the airplane, a person could see the Brocken bow only in very special circumstances. The mountaintop setting was one of the few where an observer could stand in direct sunlight with his or her antisolar point projected onto a cloud. The effect was occasionally seen next to a localized fog bank or vapor rising from a warm river or lake on a cold morning, but it must have been rare. The airplane has changed that. Whenever an airplane flies in the sunshine above clouds, we have the possibility of seeing the rings about the plane's shadow. The same effect as the Brocken bow, it is more commonly called the glory. Plate 6-8 shows the glory and the shadow of the plane. The angular size of the glory does not depend on the distance between plane and cloud, but the angular size of the plane's shadow is smaller on more distant clouds. Because the sun is not a point of light but rather a disk, 0.5 degree in angular diameter, the edges of shadows are not sharp. Between a point in the shadow where none of the sun's rays arrive (the umbra) and a point outside the shadow where rays from the entire sun's disk arrive, there is a fuzzy region illuminated by only part of the sun's disk (the penumbra). For a shadow cast on a surface near the object, the penumbra region is narrow, but if we examine the shadow farther and farther from the object, the penumbra becomes wider and wider until it extends over the whole shadow. At such distances there is no distinct, sharp shadow of the object. Plate 6-9 shows the glory about the antisolar point, on clouds so distant that the plane's shadow cannot be seen.

From looking at Plate 6-8 you can tell where I was sitting when I took the picture. If I had been sitting next to the pilot, the glory would be centered about the nose of the plane's shadow. You can see that I was sitting immediately behind the wing. The size of the glory rings depends on the size of the drops or crystals in the same way as do the corona rings. Plate 6-10 shows a noncircular glory: The smaller drops near the edge of the cloud produce a section of the circle that is larger than the rest.

In one sense the glory is now fairly well understood. A mathematical theory (Mie scattering theory) enables us to calculate the intensity variation in the glory pattern. Unfortunately, it gives us little physical insight into the process that produces the rings.[2] Nussenzveig[3] has developed another mathematical treatment, in which he attempts to identify the paths of rays that interfere to produce the diffraction rings. His model is not simple, however, and we cannot use it to predict the size of the rings without the complicated mathematical treatment. Simple models exist for all the other phenomena described in this book – models that give a physical understanding of the phenomena. I wonder if there is no simple model containing the physical essence of the explanation of the glory.

THE HEILIGENSCHEIN

A discussion of the heiligenschein perhaps belongs neither in this chapter nor in a book on sky phenomena. It is an inherently interesting effect, however; and in addition, I offer this justification for its inclusion: A number of times, after I have described the glory or specter to an audience, a listener has mentioned seeing it also about the shadow of his or her head on the ground. What the listener has seen is the heiligenschein, which needs to be explained in order to separate it from the glory.

The heiligenschein (literally, holy light) is a brightness surrounding the shadow of your head on a rough surface and is particularly noticeable when your shadow falls on grass covered with dew drops. There is no color – only a white halo about the shadow of your head. Although there are two mechanisms that give rise to the heiligenschein, the effects look the same. One mechanism requires dew droplets (producing an effect that we might call the dew heiligenschein), whereas the other requires only a rough surface (the dry heiligenschein). Let us consider the effect of the water droplets first.

DEW HEILIGENSCHEIN

Bear with me while I discuss some properties of animal eyes that, as you will see, are closely connected to the dew heiligenschein. Light rays from a distant source can enter an eye and be brought to focus on the retina as shown schematically in Figure 6-1A. If that illuminated spot on the retina acts as a secondary source of light, the light that is scattered from that spot and passes back through the front of the eye will go out in the same direction it came in, headed back toward the distant source. The eye does not act as a mirror, which reflects light in different directions depending on how the mirror is positioned; rather, it has the usual property of sending light back toward the source, *regardless of its orientation*. The retrodirection is not perfect, and so the returning light spreads out a bit, but not very much. If you shine a flashlight on a dog or cat ten meters away, you can see the eyes light up if you hold the flashlight right next to your head, but the glow disappears if you move the flashlight only a half meter or so to the side of your head. Human eyes do not show the effect as strongly as those of some other animals, but the effect is still there – to the dismay of manufacturers of small cameras. They call it the red-eye problem, and it appears in flash pictures as devilishly glowing red eyes, especially of children, when the flash bulb is mounted close to the camera lens. One solution is to mount the bulb sufficiently far from the camera so that the retroreflected light from the eyes will miss the camera lens.

Another place where this same effect can be seen is on highway signs or license plates where the surface is covered with small glass beads, each of which acts as a small eyeball, sending light selectively back in the direction of the car headlights that illuminate it (Figure 6-1B).

In the dew heiligenschein, the role of eyeball or glass bead is taken by the nearly spherical dew drops supported on a blade of grass. Water does not bend light as much as glass (it has a lower index of refraction); so it brings the light to a focus a short distance behind the drop, as illustrated in Figure 6-1C. To retrodirect the sunlight most effectively, the leaf surface should be located behind the rear surface of the drop at this focus. Some kinds of grass, having leaf surfaces covered with fine hairs, do support the drop away from the surface and send a surprisingly high intensity of light back toward the source of illumination. To see such light you look toward your antisolar point, marked by the shadow of your head. Plate 6-11 shows the effect. There is an increased light intensity from the grass, which falls off as you look farther away from the shadow of your head. As in the case of the glory or specter, each person will see the halo only about his or her own shadow.

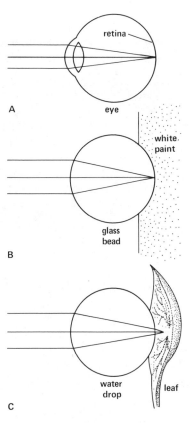

Figure 6-1. A: Light focused in an eye. B: Light focused by a glass sphere in a beaded highway sign. C: Light focused on a leaf by a drop of water.

DRY HEILIGENSCHEIN

Sometimes you can see the heiligenschein effect when your shadow falls on dry grass, or distant trees, or even a plowed field. Consider a group of leaves spread loosely through a volume of space. When the sunlight falls on this assemblage, each leaf that receives some sunlight casts a shadow on other leaves behind it. If you look directly along the path of the incident sunlight, you do not see any of the shadows because, from that direction, each shadow is directly behind its leaf. As viewed from that direction, every leaf surface you can see is illuminated by sunlight. If, however, you look off to one side, you see some of the shadows cast by the leaves. Another way of saying "Look directly along the direction of incident sunlight" is "Look in the direction of your antisolar point." The result is the same effect as before: There is a higher background intensity in the

147

grass or leaves around the shadow of your head than farther away. The effect is easier to see than to photograph; motion seems to blur the structure of the background and make it easier to see the light intensity effect. Once you tune in to it, you will see your "holy light" frequently as you walk or ride your bicycle. You will see it around the shadow of your airplane or, if you are too high above the ground to cast a shadow (owing to the angular diameter of the sun), you will see a heiligenschein spot at your antisolar point on the ground. It is even easy to see the dew heiligenschein at night from a street light.

HEILIGENSCHEIN STREAK

I have been aware since I was a child of an effect seen sometimes above the shadow of the car in which I was riding as it moved along a road with grass at the sides. There seems to be a streak of vertical light rising above the car's shadow. It surprised me as an adult to realize that this effect is a special version of the heiligenschein, which I had known about for years. It is easy to understand. Suppose, instead of the leaves arranged randomly in space, we have a collection of vertically oriented cylinders distributed loosely through space — like tree trunks in a wood. As before, if you look toward your antisolar point, every surface will be illuminated. If you look off to the left or right, you will begin to see some of the shadows cast by the vertical cylinders and, hence, a decrease in the general background brightness. However, if you look above or below your antisolar point, each cylinder still hides its own shadow and all the surfaces you see will be illuminated. Instead of a heiligenschein spot you get a heiligenschein streak. The assemblage of vertical cylinders is reasonably approximated by the stems and leaves of some grasses in certain stages of their development.

Plate 6-12 shows the streak heiligenschein seen from a small airplane flying over a field of oats or wheat. The vertical orientation of the streak surrounding the plane's shadow is quite obvious. It reveals something about the shape and orientation of the plant leaves from a distance so great that the individual leaves could not even be photographed.

I will end my discussion of heiligenschein with a quotation from Benvenuto Cellini, the sixteenth-century sculptor and metalsmith. Cellini's *Autobiography* gives insight into the political and social life of Italy in the 1500s, and throughout, it presents the vigorous picture of a man whose very considerable talent is at least matched by the size of his ego. During an imprisonment, Cellini experienced some strong religious experiences that, although changing his convictions, hardly affected his ego involvement in them. He was spir-

ited out of prison by the timely action of friends and was later to
write:

> I will not omit to relate another circumstance also, which is
> perhaps the most remarkable which has ever happened to any
> one. I do so in order to justify the divinity of God and of His
> secrets, who deigned to grant me that great favour; for ever
> since the time of my strange vision until now an aureole of
> glory (marvellous to relate) has rested on my head. This is visi-
> ble to every sort of men to whom I have chosen to point it out;
> but those have been very few. This halo can be observed above
> my shadow in the morning from the rising of the sun for about
> two hours, and far better when the grass is drenched with dew.
> It is also visible at evening about sunset. I became aware of it
> in France at Paris; for the air in those parts is so much freer
> from mist that one can see it there far better manifested than in
> Italy, mists being far more frequent among us. However, I am
> always able to see it and to show it to others, but not so well as
> in the country I have mentioned.[4]

I would say that Cellini gives a rather good description of the
heiligenschein. The only part which seems questionable is his statement that the light around his head could also be seen by those
favored few men to whom he chose to point it out. But given a style
of life that involved fatal duels over personal disagreements, such a
statement is probably understandable.

CONTRAIL AND GLORY PUZZLES

Plates 6-13 and 6-14 each show a glory, but each shows an additional feature that you should be able to explain. In Plate 6-13, what
about the dark streak extending off to the right of the glory? Plate 6-
14, which might be entitled "The Specter of the Railroad Bridge,"
shows the glory immediately around the specter's head, but what
about the larger white circle?

Plate 6-15 shows two color displays: the arc at the top of the picture and another in the contrail, right behind the airplane. Can you
explain them? Think about them before referring to the Appendix.

7

Atmospheric refraction: mirages, twinkling stars, and the green flash

On the carnival midway of the state fair there is usually a House of Mirrors. For the price of admission, a customer can wander around through a maze where some of the wall panels are mirrors and others are clear glass. In one part of the maze there are several curved mirrors that can drastically change proportions. You can see yourself tall and thin or, in another mirror, short and fat. Other mirrors lengthen one part of the body but compress another part, giving you the general appearance of either a bowling pin or a wine glass. If you can step back far enough from one of these mirrors you can even see a portion of your body detached from the rest.

There is a strong similarity between those distorted reflections and the mirages we see, usually near the earth's horizon. There is no doubt that these mirages are real physical effects: Photographs attest to their objective existence. Like the fun-house images, the mirages are distorted views of the surrounding world. In the fun house the distorting optical element is the curved mirror; in the case of the mirage it is the refracting properties of a nonuniform atmosphere. Some complicated mirages are beyond my detailed understanding, but let us begin with some simpler cases.

ATMOSPHERIC REFRACTION

Note first an effect that is always present but may not be obvious to the casual observer: As our earth rotates about its axis, the stars appear to move uniformly in circular arcs across the sky. If someone measures the angular position of a star at one time on its circular route, it is not difficult to predict where it should be at any later time. However, if the predictions and measurements take us near the horizon, we find a discrepancy between the two. The star appears to be higher above the horizon than it should be, and the discrepancy increases as the star approaches the horizon, where it is about one half of a degree above its predicted position. The explanation for the disparity is to be found in the bending (refraction) of

151

a light ray as it passes through the earth's atmosphere. To the light wave the refraction is perhaps a trivial event, taking place in the last fragment of a second of a journey from the star that may have taken thousands of years. At the bottom of the earth's atmosphere, however, our perception of the star is significantly affected. Because an understanding of this effect is essential to an understanding of mirages, a discussion in some detail is in order.

Fig. 7-1 combines two different graphic descriptions of a traveling light wave: a series of nearly parallel lines representing wave fronts and a line running perpendicular to them that represents a light ray. In the analogy with water waves, the wave fronts would represent lines running along the crests of the waves, and the ray would represent a line showing the direction in which the wave is traveling – a direction perpendicular to the crest of the wave. Light travels with its maximum speed in a vacuum; when it travels through any material medium its speed is diminished. You might then expect (correctly) that light traveling through air travels slower as the density of the air increases. That is to say, the lower light penetrates into the earth's atmosphere, the slower it travels. Consider the effect on one of the wave fronts shown in Figure 7-1. The lower part of the wave front is always lower in the atmosphere and, hence, moving slower than the upper part. The effect turns the wave front toward the earth, as shown in exaggerated scale. The light ray, which is always perpendicular to the wave front, follows a curved path and appears to the earthbound observer to be coming from a higher elevation in the sky than the actual position of the star. The effect is greatest for a star on the horizon and disappears for a star directly overhead.

It is useful in thinking about mirages to have an easy way of remembering which way a ray bends: You might think of the light ray as always bending to go into the denser air. A nice demonstration of this kind of refraction of light beam has been described by

Figure 7-1. Starlight is bent (refracted) as it passes through the earth's atmosphere, so that the star appears at a higher elevation than its actual position.

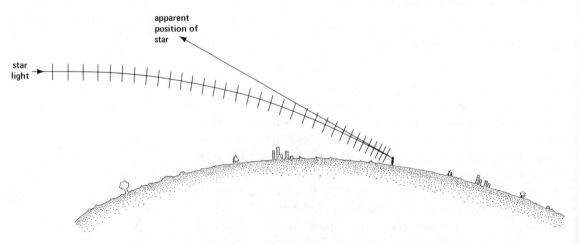

apparent
position of
star

star
light

Strouse.[1] For this demonstration sugar is put in the bottom of a glass tank filled with water. When the tank is left undisturbed, the sugar slowly dissolves, giving a solution that decreases in sugar concentration up from the bottom of the tank. Because light travels slower in a sugar solution than in pure water, a light beam (from a laser in this case) is bent, passing through the tank, in the same way a light ray is bent passing through the atmosphere. Figure 7-2 shows the obvious bending.

If navigators are fixing their position on earth by measuring the direction of a star near the horizon, they use standard tables to apply the appropriate correction for the deviation introduced by the refraction of the earth's atmosphere.

Figure 7-2. A laser beam is bent in passing through a tank containing a sugar solution of varying concentration. The beam is reflected once by a mirror at the bottom of the tank. (Photographed by William A. Strouse)

THE FLATTENED SUN

There is one obvious, naked-eye observation that can show us the effect of atmospheric refraction. When the sun is near the horizon, refraction will raise the apparent position of the top edge of the sun, but the bottom edge, nearer the horizon, will be raised even more; the result is the flattened appearance of the sun shown in Plate 7-1. The maximum refraction effect is about equal to the angular diameter of the sun; so when you see the sun sitting on the horizon, it may be, geometrically, totally below the horizon: It is like being able to see around the corner.

ATMOSPHERIC REFRACTION AND TEMPERATURE VARIATION

The atmospheric refraction under discussion arises from the density variation of the air, which results primarily from the change in pressure with altitude. Air molecules experience a gravitational pull, just as do lumps of coal, and at the bottom of the coal pile or the pile of air, the pressure is higher than at the top. Air, unlike coal, is able to compress and become more dense under pressure. However, most mirage effects take place in the lower part of the atmosphere over such a small range of altitude that the density changes owing to pressure variations are very small. For most mirage effects the temperature variations are the important factor. When air is heated it expands, with the result that, under the same pressure, hot air is less dense than cold air. To the idea of light bending toward the *denser* air we should add the realization that, when we can neglect pressure effects, light will bend so as to enter the *colder* air. To know the path of a ray through the air, then, we need to know how the air temperature varies with height.

THE DESERT MIRAGE (INFERIOR MIRAGE)

Because the word *mirage* often brings to mind a blazing sun over the hot desert sand, with waving palm trees seen next to a lake in the distance, let us begin with this particular mirage. During the day the sand becomes hotter than the air and heats the air layer right next to the surface to a higher temperature than the air a few meters above. A light ray passing near the sand will be bent upward into the cooler (denser) air. Fig. 7-3 shows several rays, leaving the same point on a palm tree, headed in different directions. The most drastic bending takes place closest to the ground in this case, where the temperature is changing most rapidly with height.

Figure 7-3. Light-ray paths for an inferior mirage. The air temperature decreases with height.

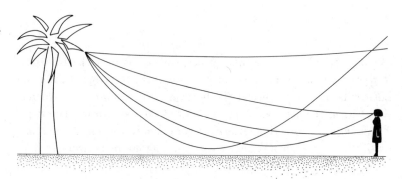

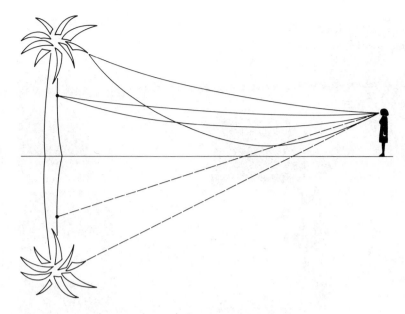

Figure 7-4. The origin of the inverted image in the desert mirage.

(These drawings greatly exaggerate the effect to make it easier to discuss.) The only rays of particular importance to the observer in the figure are those that reach her eye, two rays in this case. Figure 7-4 shows only the rays that reach her eye starting from two different points on the tree. The dotted lines show where the lower rays appear to be coming from. The upper rays give a direct view of the object (slightly lowered from its true position), whereas the lower rays give an inverted image. The only other place in nature where we see a large real object and an inverted image below it is by reflection from a water surface. When confronted, then, with the inverted picture of tree, sky, and other nearby objects, the mind interprets this as reflection off a water surface. It should be clear from the description that not only is there no water, but there is no reflection – only refraction – producing the mirror image. Both the upper and the lower image can be stretched or compressed vertically, depending on the details of the temperature variation with height.

This is the mirage seen on a hot road, the most familiar mirage for many of us (Figure 7-5). We see it as a wet place on the road ahead, without realizing that it looks wet because of the apparent reflection of more distant landscape features or of the sky. Of course, as we approach, the wet place recedes into the distance, just as the desert traveler is continually betrayed by the false promise of water just ahead.

The kind of mirage formed over a hot surface is sometimes referred to as an inferior mirage. This is not a comment on its optical or aesthetic quality but a reference to the fact that the upright mirage image appears below its real position. By contrast, in a supe-

155

rior mirage (discussed in a subsequent section), the image is seen
above the actual position of the object. Figure 7-6 brings out another important feature of the common inferior mirage. Here you
see light rays coming from the back of a monkey part way up the
tree trunk. The rays numbered 1 through 5 start off in different directions, each successive one heading downward more than the previous one. The rays are traced some distance away, to a point where
an observer might be standing, let us say, against a wall. Ray 1 arrives high on the wall at A; ray 2 arrives lower at B; and ray 3 strikes
the wall even lower at C. So far there are no surprises. But ray 4,
which starts out lower than 3, ends up higher on the wall at point B,
and ray 5 arrives even higher at A. Of all the possible rays coming
from the monkey, starting off *in any direction,* none can arrive at the
wall at any point lower than C. If I put my eye at A, I will see a
direct view of the monkey (via ray 1) and an inverted view (via ray
5). Similarly, with my eye at point B, I will see two images, but
from the angle between the rays you can see that the images will be
closer together than at A. As I move my eye down to point C, the
two images come together into one, and if I move my eye to point D
the monkey disappears! With my eye at point C, I see the top of the
tree but none of the tree trunk below the monkey.

Figure 7-4 was therefore not drawn quite correctly; the paths of

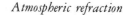

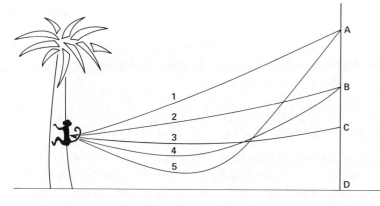

Figure 7-6. The paths of five rays under desert mirage conditions.

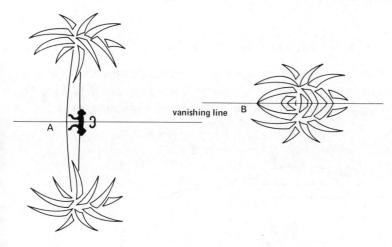

Figure 7-7. The origin of a vanishing line in the desert mirage.

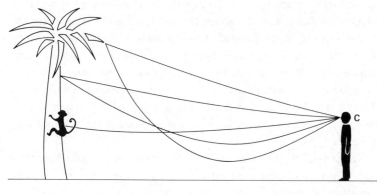

Figure 7-8. The appearance of the desert mirage scene with (A) the observer positioned as shown in Figure 7-7 and (B) the observer farther from the tree.

the rays should look like those in Figure 7-7, and the view should be that of Figure 7-8A. There is a vanishing line, which passes through the monkey's back, and I can see nothing on the tree below it. The inverted image appears to be a reflection at this vanishing line. If I look at a tree further away, the vanishing line will rise and I will see something like Figure 7-8B. For a tree yet further away, the vanishing line will be above the tree and the tree will not be visible. You can understand this distance effect from Figure 7-6. Consider my eye level to be point C. As I move away from the tree to the right, ray 3, which comes to point C, will pass over my head and I will not be able to see the monkey. I will have a new vanishing line higher in the tree. With this background, I think we can understand many of the features of inferior mirages.

ISLANDS VIEWED ACROSS WARM WATER

Consider a group of islands that, viewed across an expanse of water, look like those in Figure 7-9A in the absence of mirage conditions. If the water is warmer than the air, the appearance changes to that of Figure 7-9B, where you see the top of the islands with the inverted image apparently reflected about the vanishing line. Frequently the inverted image is compressed, looking more like that of Figure 7-9C. Figure 7-10 shows such a picture taken over Lake Erie on a cool summer day. When I stood in the same place but stooped down, changing my eye level by about a meter, the scene changed to that of Figure 7-11. You can understand the change by just assuming that the vanishing line is raised (as if you went from point C to point D in Figure 7-6).

Figure 7-12 shows a series of pictures of the setting sun; Figure 7-13 explains the shapes as an inferior mirage with an apparent reflection about a vanishing line that lies above the horizon.

INFERIOR MIRAGES ACROSS A FROZEN LAKE

Figure 7-9. A: Appearance of islands with no mirage conditions. B: Apparent reflection about a vanishing line. C: Vertical compression of inverted image.

Plate 7-2 shows a scene on a cold day in March when the air temperature was lower than the ice and water temperature of Lake Michigan. (Remember that this is the same condition that produces the

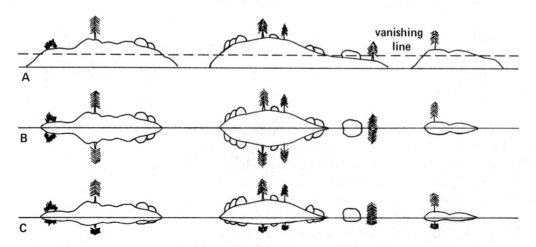

158

Figure 7-10. Islands in Lake Erie as photographed by the author standing on shore.

Figure 7-11. The islands of Figure 7-10 photographed by the author from a stooping position, which lowered the camera by about a meter.

desert mirage – where the surface is warmer than the air above.) There is a vanishing line on the bluff rising from the lake shore, and the scene of snow, rocks, and trees just above this line is seen inverted just below it. Another photograph (Plate 7-3) taken at the same time shows that the land at the end of a point is seen in apparent reflection, with the vanishing line clearly above the apparent horizon. This is a greatly magnified version of the effect shown in Figures 7-10 and 7-11. You can see that the scene of Plate 7-2, which is a common winter mirage over frozen lakes, is optically the same as the desert mirage.

SUPERIOR MIRAGES

All of the mirage effects discussed so far form above a surface that is warmer than the adjacent air. Another class of effects arises when warmer air moves over a colder lake or ocean. The air layer near the surface is cooled, causing light rays to be bent toward the earth and hence making an object appear to be raised above its true position, in a superior mirage. The simplest example is illustrated in Figure

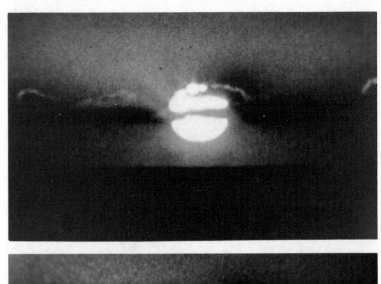

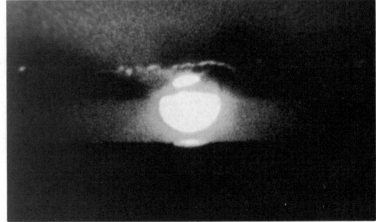

Figure 7-12. Sequence showing the setting sun. (Photographed by Gerald Rassweiler)

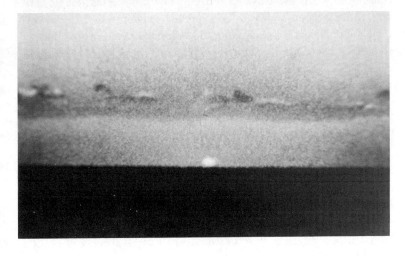

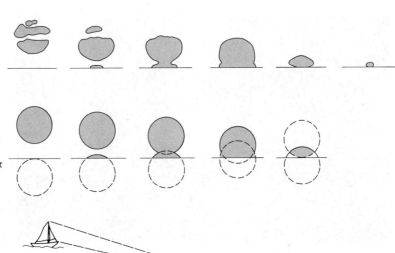

Figure 7-13. Drawing illustrating how the photographs of Figure 7-12 can be explained as an inferior mirage with an apparent reflection about a vanishing line above the horizon. (Drawn by Gerald Rassweiler)

Figure 7-14. Rays producing a superior mirage.

7-14, where the mirage image is elevated. In the absence of other effects the presence of such a mirage may not be very noticeable. Sailors have long called this condition *looming,* and when it is well developed, it can let us see a distant ship that is geometrically below the horizon, much as we can see the sun after it has set. The air temperatures may vary with height so as to produce other, more noticeable effects than the elevated position of the superior image. The noticeable image may either be stretched or compressed in the vertical dimension. Old-time sailors had descriptive terms for both of these conditions also: *towering* and *stooping.* These effects have their analogue in the thin or fat images you see of yourself in the funhouse mirrors.

In the earth's lower atmosphere, the air temperature usually decreases with increasing height. In the summertime, at 9,000 meters altitude (within cruising altitude for a commercial airliner), the outside temperature may well be around -40 degrees Celsius. A typical rate of change of temperature is 7 degrees Celsius per thousand meters of altitude. The case I have described, where the air immediately over a cold surface increases in temperature for some distance above, is called a temperature inversion. It need not start at ground level but may have the form of a layer of air some distance above the ground level that is warmer than the air either above or below it. Figure 7-15 shows how the temperature might change with height through an inversion and how light rays traveling at different heights could be bent by different amounts in different directions.

162

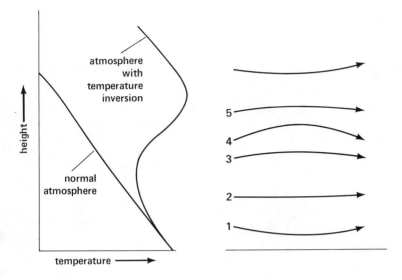

Figure 7-15. The curvature of light rays at different heights in an atmosphere with a temperature inversion.

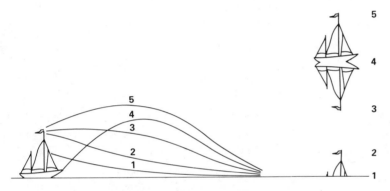

Figure 7-16. Ray paths through an atmosphere like the one illustrated in Figure 7-15. The appearance of the resulting three-part mirage is shown on the right.

A THREE-PART SUPERIOR MIRAGE

Figure 7-16 shows the path of some rays from a sailboat reaching an observer's eye after passing through an atmosphere with the temperature structure described in Figure 7-15. When ray 1 reaches the observer's eye, it is traveling in an exactly horizontal direction. Rays 2, 3, 4, and 5 come to the observer from progressively greater angles above the horizontal. The rays, over the middle part of their path between ship and ship watcher, correspond to the rays with corresponding numbers in Figure 7-15. On the right side of Figure 7-16, the numbers show the parts of the three-part mirage that would be revealed by each of the five rays.

On a hot, sultry afternoon in August 1797, the Reverend S. Vince – a fellow of the Royal Society of London and thus a dedicated amateur scientist – looked out over the North Sea from Ramsgate in southeastern England and noticed some unusual refraction effects. He examined ships through his telescope, made drawings, and later described his observations to a gathering of the Royal Society.[2] Sev-

163

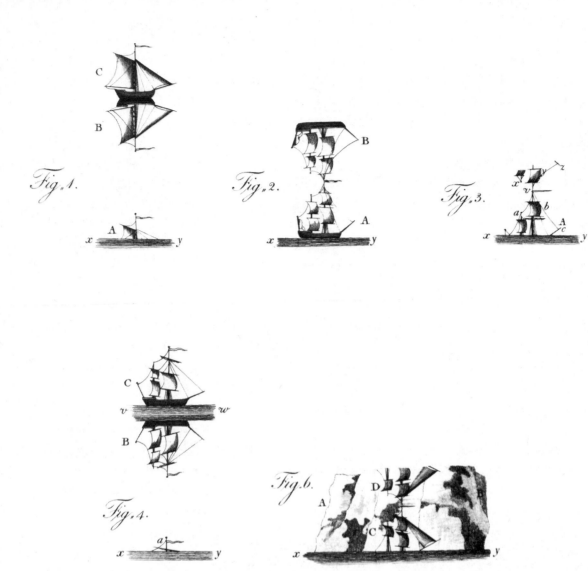

Figure 7-17. Some observations of mirages explained by Figure 7-16. (From S. Vince. "Observations on an Unusual Horizontal Refraction of the Air, with Remarks on the Variations to Which the Lower Parts of the Atmosphere Are Sometimes Subject." *Philosophical Transactions of the Royal Society of London* 89, 436 [1799])

eral of his drawings are reproduced in Figure 7-17. Two of them (his figures 1 and 4) look much like the sketch of Figure 7-16 (that is no great coincidence, of course, as I made the sketch to resemble the drawing of Mr. Vince). His description indicates that this was the appearance of distant ships. For nearer ships the upper images were lower, and in general, for the nearer ships the upper parts of the mirage image disappeared.

You might think that by seeing how the mirage appearance varies with the distance to the ship and by looking at the consequences of different temperature variations, we could understand all of Vince's drawings. In fact, several scientists are making progress on the problem of using mirages as data from which they can mathematically deduce the temperature distribution of the lower atmosphere.

The bending of light rays associated with temperature inversions are responsible for some of our most impressive mirages. An example occurred on the evening of April 26, 1977, when the residents of Grand Haven in Michigan looked west over Lake Michigan and saw city lights. The city of Milwaukee, seventy-five miles distant, is geometrically well below the horizon and out of sight from Grand Haven. An observation of a blinking red light convinced the viewers that Milwaukee was this city-above-the-lake: An ingenious observer timed the blinking and called someone in Milwaukee to find out what it could be. His measurement agreed with the frequency of the flashing red light beamed eastward from the entrance of the Milwaukee harbor. Weather Service records for that time showed a strong temperature inversion over the lake.

THE FATA MORGANA

Among the many possible mirage forms, one has a name particularly suggestive of magic and mystery: the fata morgana. Humphreys[3] traces the legend of the fata morgana in this way:

> Morgana (Breton equivalent of sea woman) according to Celtic legend and Arthurian romance, was a fairy, half-sister of King Arthur, who exhibited her powers by the mirage. Italian poets represent her as dwelling in a crystal palace beneath the waves. Hence, presumably, the name Fata Morgana (Italian for Morgan le Fay, or Morgan the fairy) was given, centuries ago, to those complicated mirages that occasionally appear over the strait of Messina . . . molding the bluffs and houses on the opposite shore into wondrous castles that, alike, tower into the sky and sink beneath the surface; nor is it strange that this poetical name should have become generic, as it has, for all such multiple mirages, whenever they occur.

For the fata morgana to occur, there must be an atmospheric condition that draws small features into vertical streaks, creating the illusion of walls and structures surmounted by spires. An observer looking at different elevations above the horizon (as along rays 1, 2, 3, and 4 in Figure 7-18) sees the same point, or nearly the same point. This can transform a fairly flat horizon into a vertical wall. Plate 7-4 is a photograph of such an effect, taken at the Naval Arctic Research Laboratory at Point Barrow, Alaska. I was looking out over the frozen Arctic Ocean, past a small laboratory building on the ice. Normally, the frozen ocean demonstrated the usual conventions of perspective as it stretched out to the distant line of its horizon. On this morning, however, the usual conventions were suspended. A wall of ice seemed to rise vertically some distance beyond the hut. When you look at the photograph it seems as if you can judge the distance out to where the wall sits on the rough sea ice, a solid feature of this frozen seascape. (Plate 7-5 shows a similar view in another direction, where the refraction had drawn the high points of the rough ice into columns and spires.) An hour later, the rough ice in the foreground was still there but the wall had disappeared, leaving behind only the results of light acting on film — a persistent chemical memory that survived the event and produced the photographs.

For these effects to occur, the temperature variation in the air must be just right to bend the rays leaving the object down to the observer's eye (Figure 7-18). If the temperature profile (the variation

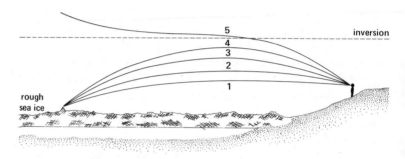

Figure 7-18. Ray paths for the fata morgana.

of air temperature with height) were such as to bring rays 1, 3, and 4 to the eye, but not ray 2, part of the wall would be missing, as in Plate 7-6. Or part of the spires could be missing, as in the delicate construction of Plate 7-7.

Of course, looking at a scene like that of Figure 7-19, you may believe that you are seeing things as they really are: just an interesting ice formation. But Figure 7-20 is the same view as 7-19 with the telescope and camera lowered by about 30 centimeters, and Figure 7-21 shows the scene after I had walked down the slope toward the frozen ocean, lowering my viewpoint an additional 120 centimeters to a point about 3 meters above the ice surface. As I moved down closer to the ice, the structure rising above the horizon disappeared altogether — disappeared, that is, until I climbed back up to look from my former position.

Fraser[4] has pointed out that in some cases it is not even necessary to have an irregular horizon to produce the fata morgana. If there are undulations in the height of the inversion layer, owing to long slow waves passing through the atmosphere, the flat sea horizon may be lifted in some places but not others, creating the appearance of walls and buildings.

Figure 7-19. View across the frozen Arctic Ocean from a point about 4.5 meters above the surface. (Photographed by the author)

Figure 7-20. The view of Figure 7-19 with the camera lowered about 30 centimeters.

Figure 7-21. The view of Figure 7-20 with the camera lowered an additional 120 centimeters.

There is another refraction effect that can produce some unusual mirage effects. Sometimes a light ray may get trapped by an atmospheric layer and kept in this layer as it travels over great distances. Figure 7-22 illustrates how it might happen. Remember the idea that light rays will be refracted in a nonuniform atmosphere so as to move toward the cooler layer of air. The height—temperature graph in Figure 7-22 represents a temperature inversion such that at the height of the dotted line, a layer of cooler air lies between warmer air both above and below. Suppose that a ray of light is introduced into this layer, traveling at a small angle above the horizontal. As it travels above the minimum-temperature line, it will be refracted back down through the layer; and as it travels below, it is refracted back upward. It may travel for a long distance, tied to this invisible temperature line by optical strings that frustrate its every attempt to escape. In fact, unless the temperature structure of the atmosphere changes — at the boundary between ocean and land, for instance — the ray will not be able to escape the layer. A similar discontinuity is needed for the ray to enter the layer with the small angular inclination required for its capture.

Rays of different inclinations may oscillate with different periods, as shown in Figure 7-22. You can see that if a mountain interrupts this inversion layer at distance D, one will see quite different points on the mountain by looking in slightly different directions from point A. The appearance of the resulting mirage depends critically on the temperature distribution and can be a complicated problem to unravel. It may be that the mirage of Plate 7-8 involves the trapping (or ducting) of light rays. The Alaska Range of mountains seen in that photograph is over one hundred miles from the Geophysical Institute building at the University of Alaska where Glenn Shaw took the photo. The view from the building is over a valley with frequent strong inversion layers; hence the resulting mirage effect.

Figure 7-22. The ducting of light rays along an inversion layer.

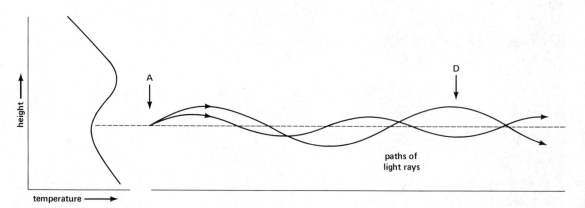

paths of
light rays

height ⟶

temperature ⟶

The ducting of light rays produces a rare effect that has been reported in polar regions. It is called the Novaya Zemlya effect because of a 1596 description of its observation on the Arctic island of that name. An expedition in search of a Northeast Passage to the Orient was frozen in over the Arctic winter at Novaya Zemlya. In the spring, two weeks before the return of the sun was expected, a flat, distorted image of the sun was visible when the sun's actual position was still about 5 degrees below the geometrical horizon. Similar observations have been reported by Antarctic investigators and have recently been given a convincing explanation by W. H. Lehn.[5] Lehn used data from an Antarctic observation that recorded measurements of the temperature variation of the lower atmosphere; with the aid of a computer, he traced rays through the atmosphere and reproduced the strange shape and size of the sun as it appeared when it was actually 4.3 degrees below the horizon. He concluded that the light rays responsible for this mirage followed the minimum-temperature line — bending with the curvature of the earth — for a distance of about 400 kilometers before they broke out of the layer.

THE ARCTIC MIRAGE IN FOLK LEGENDS

Sawatzky and Lehn[6] have examined the folk legends of the rural people living near the North and Baltic Seas and the North Atlantic Ocean and have concluded that knowledge of mirage effects was common, especially to seafaring folk, who understood that mirages could give them information about what lay beyond the horizon. Sawatzky and Lehn suggest that the original Celtic discoverers of Iceland may have ventured forth in their fragile skin-covered boats at the prompting of mirage information from the Faroe Islands, located at a distance of 385 kilometers from Iceland. From the legend of the discovery of Greenland by Erik the Red, they also suspect that Erik had some indication of the presence of land in the direction of Greenland. When he set out to find a place to live during his exile from Iceland, he headed directly to the nearest coast of Greenland, about 300 kilometers away. Under normal atmospheric conditions there is no line of sight between even the highest mountain peaks near both coasts. Sawatzky and Lehn point out that the prevailing winds and ocean currents would make traveling on this heading most difficult, and they conclude, "The logical basis of Erik's line of action was that he possessed information which well may have been transmitted by the arctic mirage."

In addition to the great variety of possible mirage effects, many of them highly complex, there are other observable effects – not commonly called mirages – that also result from light refraction in the atmosphere.

Twinkling of the stars is an effect not only described in an old children's nursery rhyme but actually seen by some modern children who are fortunate enough to be away from the lights and pollution of the cities on a clear night. The twinkling, or scintillation, of stars is closely related to the effects of "heat waves" rising from any hot surface. If you look across the top of a hot stove or radiator, or a blacktop road on a sunny day, the scene seems to shimmer – to tremble and flicker – as a result of the unevenly heated air. As noted earlier, a ray of light traveling through air of nonuniform temperature will tend to bend toward the cooler air. Above a heat source, such a tendency leads the ray on a twisting course through the turbulent air, its path constantly changing with the roiling of the air mass. You can see a similar effect if you look into a beaker or pan of water being heated from below. The effect has an analogue in the wavering distortion of an underwater scene, viewed from above the water surface, where the wandering of the light paths results from the surface refraction as affected by the movement of water waves on the surface. If you sit behind the wing on an airplane and look at the ground through the exhaust of the plane's engine, you will see things poorly. The shimmering effect in this case is so rapid, owing to the high speed of the turbulent exhaust, that you may not follow the visual fluctuations but instead may see their effects as a blur in the features on the ground below.

TWINKLING OF STARS

The atmosphere is always moving, parts of it being warmed or cooled by the earth and sun, and, to a greater or lesser degree, it always exhibits the effects of small-scale temperature inhomogeneities. If, instead of looking at an extended scene, you are looking at a distant streetlight or a star, you see the light dance about, rapidly changing its position as a result of the constant variation in the paths of the rays that come to your eye. There are, however, two other effects also involved in twinkling.

A blob of air at some instant between you and the source can act as a weakly converging lens, simultaneously bringing rays of light to your eye from several slightly different directions. The effect is an instantaneous enlarging and intensification of the source. Another blob of air, acting as a diverging lens, can diminish the amount of light entering your eye an instant later. The resulting fluctuation in the brightness and even the apparent size of the star is a second component of twinkling.

A third twinkling effect involves color changes, along with the dancing motion and erratic brightness. I may think of a blob of air that displaces the image of a star as acting like a prism. It refracts the light so that a ray comes to my eye from a direction slightly different from the actual direction of the star. But refraction of light by a prism involves two effects: deviation, whereby the ray changes direction, and dispersion, whereby different colors are deviated by different amounts. A small air prism may at one instant deviate to my eye blue light but not the rest of the spectrum. The air prism may then move so that, an instant later, the red light is reaching my eye. The rapid change of color is also a part of twinkling.

These last two effects will be easier to see if you look at a bright star near the horizon through a pair of binoculars or, better, hold the binoculars so that you look through one side with one eye only. If you rotate the front end of the binoculars in a small circle (one or two times per second), you will see the star drawn out into a circular scribble of light. It is much like the effect of "writing" words in the air with Fourth of July sparklers. Our persistence of vision lets us see a line of light rather than a moving spot. But in the case of the twinkling star, the line is not uniform; it shows bright spots and colored spots like beads on a string. By moving the binoculars, we have changed the temporal variation of brightness and color into a spatial variation. You can witness this temporal variation along the line painted across your retina by light arriving through our troubled atmosphere from a remote star.

The larger the angular dimension of a light source, the less pronounced is the twinkling. To see this effect, compare the appearance of streetlights at different distances: If the apparent dancing motion of a point source is smaller than the angular diameter of the source, the dancing will not be obvious. Aristotle, more than twenty-three centuries ago, was commenting on this effect when he observed that, compared with the twinkling stars, the planets shine with a much steadier light. Although the stars are much larger than the planets, their much greater distance from us results in their smaller angular diameter.

OTHER EFFECTS OF ATMOSPHERIC REFRACTION

Even if the earth's atmosphere were so stable that no twinkling would occur, the refractive properties of the air layer surrounding our planet could be seen. At the beginning of this chapter I explained how the light from a distant star follows a curving path through the atmosphere, always bending toward the denser air. The effect, illustrated in Figure 7-1, is that we see the apparent position of the star higher above the horizon than its actual position. But, again, the bending of blue light should be greater than the bending of red light; hence, the blue image of the star should appear higher above the horizon than the red image. Plate 7-9 shows such an effect on the appearance of Venus. On the left is a photograph taken through a reflecting telescope when the planet was 10 degrees above the horizon. It shows the predicted blue upper edge and red lower edge. In the central part of the image all of the colors overlap to produce white. On the right is a photo taken with Venus 2 or 3 degrees above the horizon (I will deal with the difference between the two photographs shortly). If the object of a photograph were to give information about the planet, it is obvious that the photo should be taken when the planet appears higher above the horizon; better yet, the photo should be taken from a mountaintop that rises above a significant fraction of the earth's atmosphere (and, indeed, most of our major astronomical observatories are located on mountaintops), or, still better, from a satellite that has escaped altogether the optical confusion of our atmosphere. The colors of Venus are not usually seen with the unaided eye, but this effect gives us some insight into a most interesting phenomenon that has been heard of by many, seen by some, and understood by a few – the green flash.

THE GREEN FLASH LEGEND

In 1882, Jules Verne published a romantic novel, set in Scotland, entitled *Le Rayon Vert*. This book brought the phenomenon of the green ray (or green flash, as it is commonly called in English writings) to the attention of the general public and was probably responsible both for stimulating much of the subsequent interest in this effect and for surrounding the green flash with something of the romantic mystique that still persists. In *Le Rayon Vert*, Miss Campbell reads this article from the *Morning Post:*

"Have you sometimes observed the sun set over the sea? Have you watched it till the upper rim of its disc, skimming the surface of the water, is just about to disappear? Very likely you have: but did you notice the phenomenon which occurs at the very instant the heavenly body sends forth its last ray, which, if the sky be cloudless, is of unparalleled purity? No perhaps not. Well the first time you have the opportunity, and it happens but rarely, of making this observation, it will not be, as one might think, a crimson ray which falls on the retina. It will be a green, but a most wonderful green, a green which no artist could ever obtain on his palette, a green of which neither the varied tints of vegetation nor the shades of the most limpid sea could ever produce the like! If there is a green in Paradise, it cannot but be of this shade, which most surely is the true green of Hope."[7]

What the article did not mention, but what Miss Campbell knew,

was that this Green Ray tallied with an ancient legend, which till now she had never been able to understand. It was one of the numerous inexplicable legends of the Highlands, which avers that this ray has the virtue of making him who has seen it impossible to be deceived in matters of sentiment; at its appearance all deceit and falsehood are done away, and he who has been fortunate enough once to behold it is enabled to see closely into his own heart and to read the thoughts of others.[8]

D. J. K. O'Connell,[9] who has reviewed the history of the green flash, comments, "It would be interesting to know what drew Verne's attention to the matter, because one can find hardly any mention of it before that time." He cites modern attempts to explain the phenomenon, including one theory that the horizontal rays of the sun, setting over the ocean, turn green by passing through water waves; and another (medical) explanation that attributed the effect to biliousness in the observer. The first explanation can be disposed of by considering that the effect has been seen over the desert; the second is, admittedly, a bit more difficult to contest.

Another, more specific, physiological explanation held that the color is an afterimage effect in the eye: Immediately after staring at the brilliant orange of the setting sun, you see as it disappears an afterimage of the complementary color — blue, or perhaps green. This theory, being more specific than the bilious theory, should be easier either to verify or to disprove. In fact, the complementary color theory was adequately dealt with by the first observer to look for — and see — the green flash on the rising sun.

UNDERSTANDING THE GREEN FLASH

Actually, the green flash is quite easy to understand, and if you are willing to count observations made through binoculars or a telescope (less spectacular, to be sure, than the description in the *Morning Post*), it is not as rare as legend would have it.

If you look at a nearby light bulb through a glass prism, held with the base down, you will see a displaced image of the white bulb fringed above with blue and below with red. You can consider the effect a whole series of overlapping images, each color forming an image slightly displaced vertically from the next. In the center, where all the images overlap, the color is white. Part of the blue image is seen by itself above the rest of the images, and part of the

173

red, below. If you were to bring a piece of cardboard up, in front of the prism, you could cut off all of the light bulb image except the upper blue edge; and by blocking out the rest of the bright white disk, you could see the blue edge more clearly.

The analogy between this situation and the conditions that produce the green flash is almost complete. The light bulb, of course, serves as the sun; the prism, as the refracting atmosphere; and the occulting cardboard, as the earth's horizon. Without the horizon, this explanation is just a repetition of the effect shown with Venus in the left photograph of Plate 7-9. The obvious difficulty with this model is that it would have us expect a blue flash rather than a green flash. But another mechanism is also operating: The atmosphere is selectively removing the blue end of the spectrum from the sunlight, over its long air path, by scattering. As the sun sets, atmospheric refraction would cause the blue image of the sun to linger and disappear below the horizon last; but atmospheric scattering is most effective in removing this blue end of the spectrum from the transmitted light. The resulting compromise is, sometimes, that the last light to be seen is green. The right photograph in Plate 7-9 shows the effect for Venus while it is still 2 or 3 degrees above the horizon. Blue-light scattering has resulted in a green upper edge on the planet's image. Under conditions of very clear atmosphere the flash has been reported as blue, and on several occasions, through a telescope, I have seen a ragged blue upper edge on the sun as it approached the horizon.

From this explanation we can deduce some of the conditions that would enhance our chances of viewing the green flash. A low horizon enables us to look at the sun where the atmosphere has the greatest dispersion — where the colors are separated most widely in passing through the atmosphere's prism. An obvious place for a low horizon is over an ocean or large lake. Also, the atmosphere must be free enough from scattering particles so that, even through the longest air path, an appreciable amount of green light is transmitted. If the sun near the horizon appears very red, you should not expect to see the effect; but if, as it sets over a low horizon, its appearance is still yellow, even though dimmed, the flash is likely to be visible. Binoculars or a telescope will enable you to see the effect far more often than with the unaided eye, but I am reluctant to mention that fact. Anyone who talks to the public or writes about looking at the sun is afraid of being responsible for either real or imagined damage to the eyes of someone who, in response to the presentation, looks at the sun. Looking at the sun through binoculars or a telescope is a *very real danger* and should never be done except under circumstances where the sun's light is significantly diminished. The last moments before the sun disappears below the horizon are such a time, and the view through binoculars can show a wealth of interesting detail not to be seen otherwise.

Glenn Shaw[10] at the University of Alaska has done some quantitative predictions concerning the model of the green flash discussed here. He considers the amount of refraction and dispersion expected from a normal atmosphere (no temperature inversion) and evaluates the light scattered by two mechanisms, the blue-sky (Rayleigh) scattering from the air itself and the scattering resulting from small aerosol particles (solid or liquid particles in the atmosphere) of a type and concentration found in a relatively clear Arctic atmosphere. His calculations predict that, as a result of these three factors, the upper edge of the sun at sunset should indeed appear green. He concludes:

It is seen that the blue-green to yellow-green region extends over a vertical angular distance of approximately 0.15 milliradians (about 0.5 minutes of arc). Strangely enough, the topmost blue-appearing region is spatially small with respect to the green hue, and, in addition is reduced in intensity by several orders of magnitude and hence would not be apparent — a comforting fact, since qualitative arguments would lead one to expect a blue flash.

But the explanation is not complete. Our model predicts a green flash, and yet we do not see it every time the sun sets over a low horizon with a reasonably clear sky. Fraser[11] has pointed out that the angular extent of the green rim, as calculated by Shaw, is less than the minimum angle that can be resolved by the human eye. He suggests two circumstances in which the effect can become visible to the unaided eye. If the rest of the sun's disk is blocked off (by the horizon, for example), all the light you see is green and it does not matter that you cannot resolve the angular extent of the green-light source. This would seem to describe the green ray of Jules Verne's novel. Another circumstance is one in which the green segment is magnified by some mirage condition. This suggestion opens up a whole host of possible mechanisms.

THE GREEN FLASH AND DISTORTION OF THE SETTING SUN

Look at the sequence of photographs in Plate 7-10. The shape of the sun is greatly distorted by atmospheric refraction. The photos that show the sun with straight sides show it with the same horizontal diameter seen in the first picture. In general, the atmosphere is layered so that the only distortions that occur are in the vertical direction: The square portion of the sun, then, must be a vertically magnified portion of the central part of the sun, with a width equal

Figure 7-23. Setting sun showing spikes caused by waves on an inversion layer. (Photographed in Alaska by the author)

to the diameter of the sun. In the middle photos, the upper, curved rim is demagnified vertically and so flattened. Fraser[12] has showed that the presence of waves traveling along an inversion layer (toward or away from an observer) can give rise to the spikes seen on each side of the sun in the last frames of Plate 7-10 and in Figure 7-23. Each pair of spikes can be caused by one undulation in the height of the inversion layer and can be considered a mirage image of a horizontal strip across the sun, raised up to an apparent higher position. One of the most frequent manifestations of the green flash is associated with these spikes. As the spikes move up through the disk of the sun (as a result both of the sinking of the sun and of the traveling of the waves), they can pinch off – isolate – a segment of the sun from the main disk. This segment comes only from the top edge of the sun, and as it shrinks in size, it may turn green (or blue) just before it disappears. The green color typically lasts only about a second. It is difficult to photograph, as it requires not only a telescope (or long telephoto lens) and good timing but also the correct exposure, which is very difficult to determine next to the bright disk of the sun. Most photographs that have been obtained do not reproduce the vividness of the actual observation. Plate 7-11 shows a sequence that captures some green edges on the sun as it sets, as well as the blue green moment for the disappearing segment in the last photograph.

This talk of waves on an inversion layer and multiple-mirage images does not mean that our simple explanation of the green flash is wrong. The insight of the simple picture is right, but the situation is modified by layers in the atmosphere that can distort the apparent shape of the sun and, in so doing, magnify, demagnify, or distort in many ways its upper green rim. Some people have attempted to make distinctions among the various green manifestations that arise from the different distortions, with names including the green flash, green ray, green rim, and green segment. All are variations of the same effect and all, in my judgment, may appropriately be given the same name.

THE RED FLASH

Given the situation that produces a green flash, an obscuration of all of the sun's disk except for the extreme lower edge should produce a red flash. This flash can be seen as the sinking sun reappears from beneath an obscuring cloud; you can see an indication of the effect in the first photograph of Plate 7-11. It is not as dramatic as its green counterpart, because we are accustomed to seeing parts of the sun appear quite red from scattering effects.

"THE DULL CATALOGUE OF COMMON THINGS"

One of the themes of this book has been the relationship between understanding and beauty. For me, the technical explanation of the green flash detracts nothing from the charm of the Scottish legend, just as the symphony remains beautiful even when we understand the interplay of the various instruments and have some insight into the techniques of the composer. I hope you agree.

Appendix: Answers to puzzles

CHAPTER 1. A RAINBOW PUZZLE

What is strange about the photograph of Plate 1-13? Are the shadows in the wrong direction? No, the shadows of the trees define lines that appear to converge on the horizon directly above the center of the rainbow, just as you should expect for the perspective of parallel lines. Perhaps you noted that there are no clouds in the sky. That is a bit strange, although sometimes a thin veil of rain from an overhead cloud can produce a rainbow.

Consider the time of year at which the picture was taken. The bare trees covered with ice and the snow on the ground indicate clearly that it is wintertime in a northern climate. In such conditions, however, we normally do not see rainbows. The correct optical element to produce a rainbow is a sphere, the shape assumed by a falling drop of water. When temperatures are below freezing the spherical water drops are replaced by ice crystals, which are more complicated shapes. Small water droplets can sometimes remain liquid at below-freezing temperatures (this is water in a supercooled state), but seldom do drops large enough to produce a colored rainbow remain supercooled, without freezing.

So there is the question: How can there be a rainbow in the wintertime? To see the question behind the picture requires seeing with the mind as well as the eye. Frequently the recognition of an interesting question requires more awareness than does its answering, and this is a case in which the question is more interesting than the solution. I took the picture next to Niagara Falls, where spray from the falls, blowing through the trees, had not had time enough to freeze.

CHAPTER 2. TWO HALO PUZZLES

On December 21, the longest day of summer in the southern hemisphere, the south end of the earth's rotation axis is tilted 23.4

degrees toward the sun. At that time a person at the South Pole sees the sun travel, over the course of a day, completely around the sky at a constant elevation of 23.4 degrees above the horizon. An observer a few degrees of latitude away from the Pole would see the sun circle the sky, rising to a maximum elevation of slightly above 23.4 degrees and dropping to a minimum elevation of slightly below 23.4 degrees. Hence workers in Antarctica in the summertime frequently see the sun at an elevation of less than 25 degrees. This is just a little more than a halo's radius above the horizon, and only the top part of the lower tangent arc can be seen. This is the scene of Plate 2-24. If more of the arc were visible it might be easier to identify, but the bright spot of light right at the horizon has been a puzzle to a number of Antarctic observers.

Chapter 2 is devoted to the effects produced by sunlight passing through ice crystals floating in the air. But, like gold, ice crystals are where you find them. In Plate 2-25, they are in the form of frost growing on the plants in the field. The portion of the 22-degree halo formed by the sunlight on the frost gives evidence of the hexagonal nature of the frost crystals.

I occasionally see a portion of a halo formed in a layer of snow on the ground, but it is invariably a portion of the 46-degree halo — checked roughly by seeing that the arc lies about two handspans away from the sun. I have not yet been able to examine with a magnifier one of the actual crystals producing a spot of light in this halo, but it would seem possible to do so.

CHAPTER 3. PUZZLES INVOLVING REFLECTION

In Figure 3-37 you see two circles in the sky that are not concentric. The 22-degree halo around the moon causes no problem, but you are not accustomed to thinking of the parhelic circle as a circle of comparable size to the 22-degree halo. For the moon elevation of 63 degrees in this photograph, the parhelic circle has an angular radius of 27 degrees. If we are to be precise with language, I suppose we should refer to the circle about the moon as the paraselenic circle (all the names that involve the Greek, *helios,* for the sun have their counterpart using *selene* for the moon; for example, *paraselene, selenic arcs,* etc.). That does, however, seem to me to be an unnecessary complication. The sharpness of the circle is impressive. Taking account of the angular diameter of the moon, I calculate from the width of the ring that the reflecting surfaces are vertical to better than ±0.2 degrees. I see this effect also in other photographs, and I suspect that the reflection is from the end faces of pencil crystals with long axes nearly horizontal. If that were the case, there should also be a circumscribed halo present. Did you miss it? Look again and see that you can detect it, separate from the 22-degree halo at the sides.

Bartley L. Cardon has described and analyzed this photograph in "An Unusual Lunar Halo," *American Journal of Physics* 45, 331 (1977).

In Plate 3-14 you can see the bottom edge of the 22-degree halo at the top of the photograph, just below the cloud horizon; so you know that the sun is approximately 20 degrees above the horizon. The lower arc seems to look more like a Parry arc than the tangent arc (by comparison with the simulation of Figure 2-10). The subsun is prominent and appears to extend up into the Parry arc. On each side of the subsun you see the 22-degree subparhelia.

The sky effects shown in Plate 3-14 present an interesting contrast to most of the others under discussion: Whereas phenomena visible from the ground have surely been seen since before recorded time, most of the effects in this photograph could have been seen by few people before the era of the airplane.

CHAPTER 4. COMPLEX HALO PUZZLE

The delicacy of the slide transparency for Plate 4-3 leads me to suspect that some faint effects will not be reproduced on the printed page. The 22-degree halo is there, with the strong upper tangent arc. It looks as if there may also be a superimposed upper suncave Parry arc, but this is not obvious. The sun dogs are clear, and on the left side there may be a Lowitz arc present. The parhelic circle is present all the way around the sky, with both 120-degree parhelia evident. I can see a patch of brightness above the 22-degree halo where the top of the 46-degree halo or the supralateral arc should be, but it is not well defined. On the left side of the sun there is a trace of the infralateral arc. That is a rather good collection of effects from one picture. However, with close inspection I can see one more, a pair of anthelic arcs that extend back from the anthelic point almost to the upper tangent arc.

CHAPTER 5. LIGHT-SCATTERING PUZZLE

When I was first organizing a talk about light scattering in the atmosphere, I wanted to include a discussion of anticrepuscular rays. Only after I had decided that I had no photographs of the effect in question did I recognize, in the photographs of Plates 5-19 and 5-20, the illustrations I had been seeking. I was aware that sometimes rainbows show radial structure like the dark line running along a radius of the bow in Plate 5-19, or the bright segment inside the bow in Plate 5-20. I had never questioned why the structure had that shape; but, with thoughts of crepuscular rays stored in my near subconscious, the answer came along with the question. The dark streak along the radius of the bow in Plate 5-19 is a crepuscular ray headed for the antisolar point, which is the center of the rainbow. It may result from a cloud somewhere far on the other side of the sky. In Plate 5-20 the sunlight falling on the rain drops to produce the upper part of the bow is partially diminished by light clouds somewhere in the sky, but the lower, bright part of the bow is formed in clear sunlight. The edge of the partial shadow can be traced inside the bow, heading straight for its center.

CHAPTER 6. CONTRAIL AND GLORY PUZZLES

When a jet airplane passes overhead, you see its path marked with a white contrail. The contrail (short for condensation trail) results from the condensation of water vapor, produced by the fuel combustion process, as it cools in the low-temperature air at the flight altitude. When you can see the glory from an airplane, the plane must be in sunlight, casting its shadow on a cloud below. Because you have no rearview mirror outside your airplane window, you are not aware of the contrail your plane is leaving. However, you can see the contrail shadow in Plate 6-13, beginning at the plane's shadow, which is located at the center of the glory.

The wide-angle (fisheye) picture of Plate 6-14 shows the glory around the photographer's antisolar point. The large white circle is the white rainbow discussed in Chapter 1 — a circle centered on the antisolar point.

In Plate 6-15, the arc at the top of the picture is the upper tangent arc to the 22-degree halo, with perhaps a bit of the upper suncave Parry arc also showing. By comparison with the simulation of Figure 2-10, it appears that the sun elevation must be in the range of 30 to 40 degrees. Look closely at the colors in the contrail just behind the plane. In the contrail nearest the sun, you can see blue color, changing with the usual spectral sequence to red farthest from the sun. Refraction halos and arcs have their inner edges red; it is diffraction effects that put the blue light nearer to the sun than the red light. So in that contrail we are seeing a thin slice through a diffraction ring (part of the corona display) around the sun. In this case, the water droplets probably have a uniform size; hence the display of spectral colors. Kenneth Sassen has described and analyzed this photo in "Iridescence in an Aircraft Contrail," *Journal of the Optical Society of America* 69, 1080 (1979).

Notes

CHAPTER 1. RAINBOWS

1 Sir Isaac Newton, *Opticks; or, a Treatise of the Reflections, Refractions, Inflections and Colors of Light*, 4th ed. (London, 1730; reprint ed., New York, 1952), p. 178.
2 Carl B. Boyer, *The Rainbow from Myth to Mathematics* (Yoseloff, New York, 1959), pp. 249–50.
3 Jearl D. Walker, "Multiple Rainbows for Single Drops of Water and Other Liquids," *American Journal of Physics 44*, 421 (1976).
4 Alistair B. Fraser, "Inhomogeneities in the Color and Intensity of the Rainbow," *Journal of Atmospheric Sciences 29*, 211 (1972).
5 Boyer, *Rainbow*, p. 274.
6 See, for example, the photograph in H. Moysés Nussenzveig, "The Theory of the Rainbow," *Scientific American 236*, 116 (1977).
7 G. David Scott, "The Swimmer's Twin Rainbow," *American Journal of Physics 43*, 460 (1975).
8 John Harsch and Jearl D. Walker, "Double Rainbow and Dark Band in Searchlight Beam," *American Journal of Physics 43*, 453 (1975).
9 Photograph by William Sager in *Sky and Telescope 59*, 177 (1980).
10 For further details, see Robert G. Greenler, "Infrared Rainbow," *Science 173*, 1231 (1971).
11 Nussenzveig, "Theory."

CHAPTER 2. ICE-CRYSTAL REFRACTION EFFECTS

1 C. Knight and N. Knight, "Snow Crystals," *Scientific American 228*, 100 (1973).
2 J. M. Pernter and F. M. Exner, *Meteorologische Optik*, 2nd ed. (Wilhelm Braumüller, Vienna and Leipzig, 1922).
3 Alfred Wegener, "Theorie der Haupthalos," *Archiv der Deutschen Seewarte 43*, no. 2 (1926).
4 Our original results for the simulations of both the circumscribed-halo and the Parry arcs are to be found in Robert G. Greenler and A. James Mallman, "Circumscribed Halos," *Science 176*, 128 (1972); and Robert G. Greenler, A. James Mallman, James R. Mueller, and Rick Romito, "Form and Origin of the Parry Arcs, *Science 195*, 360 (1977).
5 W. E. Parry, *Journal of a Voyage for the Discovery of a Northwest Passage* (Murray, London, 1821; reprint ed., Greenwood, New York, 1968).
6 R. A. R. Tricker, *Introduction to Meteorological Optics* (American Elsevier, New York, 1970).

7   See n. 4 to this chapter.
8   E. C. W. Goldie, "A Graphical Guide to Haloes," *Weather 26*, 391 (1971); E. C. W. Goldie and J. M. Heighes, "The Berkshire Halo Display of 11 May, 1965," *Weather 23*, 61 (1968).
9   P. Putnins, "Der Bogen von Parry und andere Berührungsbogen des gewöhnlichen Ringes," *Meteorologische Zeitschrift 51*, 321 (1934).
10 K. O. L. F. Jayaweera and G. Wendler, "Lower Parry Arc of the Sun," *Weather 27*, 50 (1972).
11 Tobias Lowitz, "Déscription d'un météore remarquable, observé à St. Pétersbourg le 18 Juin 1790," *Nova Acta Academiae Scientiarum Imperialis Petropolitanae 8*, 384 (1794).
12 R. A. R. Tricker, "Observations on Certain Features to Be Seen in a Photograph of Haloes Taken by Dr. Emil Schulthess in Antarctica," *Quarterly Journal of the Royal Meteorological Society 98*, 542 (1972).
13 R. D. Edge, "Bernoulli and the Paper Dirigible," *American Journal of Physics 44*, 780 (1976).
14 Tricker, "Observations"; James R. Mueller, Robert G. Greenler, and A. James Mallmann, "The Arcs of Lowitz," *Journal of the Optical Society of America 69*, 1103 (1979).
15 Our original work on the 46-degree halo is reported in Robert G. Greenler, James R. Mueller, Werner Hahn, and A. James Mallmann, "The 46° Halo and Its Arcs," *Science 206*, 643 (1979).
16 Edgar Everhart, "The Solar Halo Complex on September 17, 1960," *Sky and Telescope 21*, 14 (1961).
17 Alistair B. Fraser, "What Size of Crystals Produce the Haloes?" *Journal of the Optical Society of America 69*, 1112 (1979).
18 E. C. Goldie, G. F. Meaden, and R. White, "The Concentric Halo Display of 14 April 1974," *Weather 31*, 304 (1976).
19 R. A. R. Tricker, "Arcs Associated with Halos of Unusual Radii," *Journal of the Optical Society of America 69*, 1093 (1979).
20 Louis Besson, "Concerning Haloes of Abnormal Radii," *Monthly Weather Review 51*, 254 (1923).

CHAPTER 3. ICE-CRYSTAL REFLECTION EFFECTS

1   M. Minnaert, *The Nature of Light and Colour in the Open Air*, trans. H. M. Kremer-Priest, rev. K. E. Brian Jay (Dover, New York, 1954), p. 202.
2   Our sun pillar results were originally described in Robert G. Greenler, Monte Drinkwine, A. James Mallmann, and George Blumenthal, "The Origin of Sun Pillars," *American Scientist 60*, 292 (1972).
3   Kenneth Sassen, "Light Pillar Climatology and Microphysics at Laramie, Wyoming," *Weatherwise*, forthcoming.
4   Plate 3-10 is discussed in Edgar Everhart, "The Solar Halo Complex of September 17, 1960," *Sky and Telescope 21*, 14 (1961); Plates 3-10 and 3-19 are both discussed in A. James Mallmann and Robert G. Greenler, "Origins of Anthelic Arcs, the Anthelic Pillar, and the Anthelion," *Journal of the Optical Society of America 69*, 1103 (1979).
5   Our initial description of the anthelic arc simulations was published in Mallmann and Greenler, "Origins of Anthelic Arcs."
6   C. S. Hastings, "A General Theory of Halos," *Monthly Weather Review 48*, 322 (1920).
7   Alfred Wegener, "Theorie der Haupthalos," *Archiv der Deutschen Seewarte 43*, no. 2 (1926).

8   The rays producing the Hastings anthelic arcs in Figure 3-21 are the rays whose paths are like those labeled type 1 and type 3 in Figure 2-14. Type 2 and type 4 rays should also contribute to the Hastings arcs for some elevations of the sun. For all the cases shown in Figure 3-21, those contributions fall outside the areas covered by the simulations.

9   R. A. R. Tricker, "A Simple Theory of Certain Helical and Anthelic Halo Arcs: The Long Hexagonal Ice Prism as a Kaleidoscope," *Quarterly Journal of the Royal Meteorological Society 99*, 649 (1973).

10 See, for example, the Ellendale display, reproduced in Figure 4-4 of the next chapter.

11 Hastings, "General Theory."

12 David Lynch and Pt. Schwartz, "Origin of the Anthelion," *Journal of the Optical Society of America 69*, 383 (1979).

13 Ibid.

14 Purchased from Spiratone, Inc., New York, N.Y.

15 The extension of these arcs to the other side of the sky is discussed later in this chapter in connection with Hevel's halo and in Chapter 4 in connection with Blake's drawing (Figure 4-5) and Hogan's photograph (Plate 4-2).

16 W. J. Humphreys, *Physics of the Air,* 3rd ed. (1940; reprint ed., Dover, New York, 1964), pp. 539 ff.

17 This is the ray path shown in Figure 3-24C that produces the anthelic pillar and anthelion of Figure 3-25 – except for one difference. Here we are considering internal reflections from a pencil crystal with a Parry-arc orientation; i.e., a reflection off a side face tilted 30 degrees from the vertical. In situations discussed previously, the pencil crystal could take any rotational orientation around its long horizontal axis.

18 R. A. R. Tricker, "The Kern Arc," *Weather 26*, 315 (1971).

19 Humphreys, *Physics,* p. 542.

20 Johannes Hevelius, *Mercurius in sole visus . . .* (S. Reiniger, 1662).

CHAPTER 4. COMPLEX DISPLAYS

1   Tobias Lowitz, "Déscription d'un météore remarquable, observé à St. Pétersbourg le 18 Juin 1790," *Nova Acta Academiae Scientiarum Imperialis Petropolitanae 8*, 384 (1794).

2   Johannes Hevelius, *Mercurius in sole visus . . .* (S. Reininger, 1662).

3   Lowitz, "Déscription."

4   W. E. Parry, *Journal of a Voyage for the Discovery of a Northwest Passage* (Murray, London, 1821; reprint ed., Greenwood, New York, 1968).

5   Frank J. Bavendick, "Beautiful Halo Display Observed at Ellendale N. Dak.," *Monthly Weather Review 48*, 330 (1920).

6   J. R. Blake, *Solar Halos in Antarctica: Australian National Antarctic Research Expedition Report No. 59, Ser. A, vol. 4, Glaciology* (Antarctic Division, Melbourne, 1961).

7   The identification of the four numbered arcs in Blake's drawing follows the suggestion of R. A. R. Tricker, "A Simple Theory of Certain Helical and Anthelic Halo Arcs: The Long Hexagonal Ice Prism as a Kaleidoscope," *Quarterly Journal of the Royal Meteorological Society 99*, 649 (1973).

CHAPTER 5. SCATTERING

1   A. J. W. Catchpole and D. W. Moodie, "Multiple Reflections in Arctic Regions," *Weather 26*, 157 (1971).

2 V. Stefansson, *The Friendly Arctic* (Macmillan, New York, 1921), p. 220.

3 W. Scoresby, *An Account of the Arctic Regions, with a History of the Northern Whale-Fishery,* vol. 1, *The Arctic* (Archibald Constable, Edinburgh, 1820; reprint ed., David and Charles, Newton Abbot, Devon, 1969), pp. 299–300.

CHAPTER 6. DIFFRACTION

1 Douglas E. Archibald, "The Large Corona Round the Sun and Moon in 1883–4–5, Generally Known as 'Bishop's Ring,' " pt. IV, sec. 1(E) of the report of the Krakatoa Committee of the Royal Society, *The Eruption of Krakatoa and Subsequent Phenomena* (Trübner, London, 1888).

2 An attempt at a physical interpretation of the results of Mie scattering theory is given by H. C. Bryant and A. J. Cox, "Mie Theory and the Glory," *Journal of the Optical Society of America 56,* 1529 (1966). For a general description of the state of understanding of the glory, see H. C. Bryant and N. Jarmie, "The Glory," *Scientific American 231,* 60 (1974).

3 H. M. Nussenzveig, "Complex Angular Momentum Theory of the Rainbow and the Glory," *Journal of the Optical Society of America 69,* 1068 (1979).

4 *Autobiography of Benvenuto Cellini,* trans. John Addington Symonds (Modern Library, New York, 1927), pp. 273–4.

CHAPTER 7. ATMOSPHERIC REFRACTION

1 William A. Strouse, "Bouncing Light Beam, *American Journal of Physics 40,* 913 (1972).

2 S. Vince, "Observations on an Unusual Horizontal Refraction of the Air, with Remarks on the Variations to Which the Lower Parts of the Atmosphere Are Sometimes Subject," *Philosophical Transactions of the Royal Society of London 89,* 436 (1799).

3 W. J. Humphreys, *Physics of the Air* 3rd ed. (1940; reprint ed., Dover, New York, 1964), pp. 474–5.

4 Alistair B. Fraser and William H. Mach, "Mirages," *Scientific American 234,* 102 (1976).

5 W. H. Lehn, "The Novaya Zemlya Effect: An Arctic Mirage," *Journal of the Optical Society of America 69,* 776 (1979).

6 H. L. Sawatzky and W. H. Lehn, "The Arctic Mirage and the Early North Atlantic," *Science 192,* 1300 (1976).

7 Jules Verne, *The Green Ray* (Associated Booksellers, Westport, Conn., 1965), p. 21.

8 Ibid.

9 D. J. K. O'Connell, "The Green Flash," *Scientific American 202,* 113 (1960). See also O'Connell's book, *The Green Flash and Other Low Sun Phenomena* (North Holland, Amsterdam, 1958).

10 Glenn E. Shaw, "Observations and Theoretical Reconstruction of the Green Flash," *Pure and Applied Geophysics 102,* 223 (1973).

11 Alistair B. Fraser, "The Green Flash and Clear Air Turbulence," *Atmosphere 13,* 1 (1975).

12 Ibid.

General references

I have not attempted to give complete documentation for every topic covered in this book. Much more complete lists of references can be found in some of the works cited in the notes for each chapter. Of the following works, the third and fourth are concerned with ice-crystal forms; the others may be consulted as good general references.

Dietze, Gerhard. *Einführung in die Optik der Atmosphaere.* Akademische Verlagsgesellschaft, Leipzig, 1957.

Humphreys, W. J. *Physics of the Air.* 3rd ed. 1940. Reprint. Dover, New York, 1964.

Knight, C. and Knight, N. "Snow Crystals." *Scientific American 228,* 100 (1973).

LaChapelle, Edward R. *Field Guide to Snow Crystals.* University of Washington Press, Seattle, 1969.

Minnaert, M. *The Nature of Light and Colour in the Open Air.* Trans. by H. M. Kremer-Priest. Rev. by K. E. Brian Jay. Dover, New York, 1954.

Tricker, R. A. R. *Introduction to Meteorological Optics.* American Elsevier, New York, 1970.

Ice Crystal Haloes. Optical Society of America, Washington, D.C., 1979.

Index